THE UNIVERSE

GEOPHYSICS AND ASTROPHYSICS MONOGRAPHS

AN INTERNATIONAL SERIES OF FUNDAMENTAL TEXTBOOKS

VOLUME 11

THE UNIVERSE

by

JOSIP KLECZEK

*Astronomical Institute,
Czechoslovak Academy of Sciences,
Ondřejov, Czechoslovakia*

D. REIDEL PUBLISHING COMPANY

DORDRECHT-HOLLAND/BOSTON-U.S.A.

Library of Congress Cataloging in Publication Data

Kleczek, Josip.
   The universe.

   (Geophysics and astrophysics monographs; v. 11)
   Bibliography: p.
   Includes index.
   1. Cosmology. I. Title. II. Series.
QB981.K64   523.1'2   76-19007
ISBN-13:978-94-010-1487-8   e-ISBN-13:978-94-010-1485-4
DOI: 10.1007/978-94-010-1485-4

Published by D. Reidel Publishing Company, P.O. Box 17, Dordrecht, Holland

Sold and distributed in the U.S.A., Canada, and Mexico
by D. Reidel Publishing Company, Inc.
Lincoln Building, 160 Old Derby Street, Hingham,
Mass. 02043, U.S.A.

TABLE OF CONTENTS

PREFACE

The universe is the largest system of all. It consists of elementary particles bound together by gravitational, electromagnetic and nuclear forces. Its structural hierarchy in space (from atomic nuclei to supergalaxies) and its evolutionary sequence (from the fireball to the diversity of present forms) is governed by the properties of elementary particles and their interactions. This book is an attempt to interpret the structure and evolution of the universe in terms of elementary particles and of their interactions.

This book is intended to present a background for students in astronomy and related sciences, such as geophysics, meteorology, plasma physics, chemistry, nuclear physics, space sciences and some others. The universe forms a general framework for all the phenomena studied by these sciences.

It was possible to squeeze an extensive range of topics from various disciplines into one book of acceptable size only under some severe limitations: (a) no references are given; (b) arguments are shortcut; (c) quantities are often expressed in the order of magnitude; and (d) formulae have been limited to a minimum. Often more hypotheses or theories exist for a phenomenon. We have chosen only one. The preference for a theory or hypothesis may be personal and the theory itself may later prove incorrect. But, many theories about a particular phenomenon would cover many pages and might lead to confusing effects.

The author is deeply obliged to Professors C. Jaschek, Zd. Kopal, B. Onderlička, A. Slettebak, to Drs Zd. Pokorný and J. Sykes for their kind perusal of the text. His sincere thanks are also due to Drs J. Bičák, J. Fischer, S. Kříž, P. Mayer, Zd. Mikulášek, and Professor Vl. Vanýsek for their comments, to Dr P. Příhoda and Mr T. Vaněk who prepared the figures and to those observatories and institutes who kindly sent their photographs.

JOSIP KLECZEK

Ondřejov, March 1975

CHAPTER 1

ELEMENTARY PARTICLES

The matter of our universe consists of elementary particles. Bodies – whether it be our own body or a star – are systems of elementary particles differing in their number and degree of organization. The existence of elementary particles – also called fundamental particles – must therefore be felt in all phenomena in the universe. The physics of elementary particles offers a deeper understanding and a more profound insight into the structure and evolution of individual bodies like atoms, molecules, crystals, rocks, planets, stars, stellar systems, and the whole universe. That is why the study of elementary particles is of fundamental importance for contemporary physics in general and for astrophysics in particular.

1.1. Properties

We can never see elementary particles and never hope to describe them in a familiar way. We can only see the tracks they leave behind – tiny black clumps of silver in a photographic emulsion (Figure 1.5), bubbles of gas in rapidly expanded liquid hydrogen (Figure 1.1) a brief flash of light in a scintillator or a spark in a spark chamber. Enormous quantities of experimental material have been accumulated and interpreted theoretically in terms of particle properties.

Some of the properties are extensions of concepts from classical mechanics (e.g. mass, energy, electrical charge), while others stem from relativistic mechanics (proper time, proper length). Modern theory (quantum mechanics) had to introduce supplementary concepts (like spin, leptonic charge, baryonic charge, isotopic spin, strangeness, parity, quantum of action, annihilation, pair production, Pauli exclusion, wave-particle duality) to describe properties and behavior of elementary particles.

A set of numbers characterizes each particle, differentiates it from the other particles, and describes its properties. Some properties are constant and characteristic of the particle (rest mass, electric charge, spin, leptonic charge, baryonic charge, strangeness, isotopic spin, parity) while others are related to the surrounding world (momentum, angular momentum, total energy).

1.1.1. MASS

Mass is the quantity of matter in a body. It determines the magnitude of gravitational force (e.g. its weight) and is a measure of resistance to acceleration by any force (that is to say inertia). The mass of elementary particles is very small. Therefore they may be easily accelerated to velocities much higher than those we know from practical experience with relatively much more massive bodies. Particles with zero mass (such as photons) move with the highest velocity, i.e. the velocity of light, immediately after their birth.

The lightest particle with non-zero mass is the electron. Its mass, $m_e = 9 \times 10^{-28}$ g, is often used as the unit for measuring masses of all the other particles. The mass of the proton is $m_p = 1836\ m_e$, and that of the neutron, $m_n = 1838.6\ m_e$. Masses and other characteristic values for elementary particles may be found in Figure 1.3.

1.1.2. ENERGY

The concept of energy has been used since about 1700 when it was called *facultas agendi* which means capacity of a body for doing work. It may have different forms such as kinetic, thermal, electric, chemical, gravitational, nuclear, radiative and rest energy of mass. Its total amount remains the same although it changes its form like an actor on stage. Due to its great variability it governs the whole hierarchy of the universe, determines structure in space and evolution in time of all bodies and systems, from elementary particles up to clusters of galaxies. The great diversity of energy transformation may be reduced however to a few interactions of elementary particles as will be discussed in Chapter 2.

A particle with rest mass m_0 moving with velocity v in a reference system (e.g. the walls of a laboratory, satellite, Sun, center of Galaxy) has an energy

$$E = \frac{m_0 c^2}{\sqrt{1 - \dfrac{v^2}{c^2}}} \tag{1.1}$$

For small velocities Equation (1.1) reduces to

$$E \approx m_0 c^2 + \tfrac{1}{2} m_0 v^2 \tag{1.2}$$

and at rest

$$E = m_0 c^2. \tag{1.3}$$

The term $m_0 c^2$ is the rest energy and $\tfrac{1}{2} m_0 v^2$ in Equation (1.2) is the kinetic energy of the moving particle. A particle with kinetic energy comparable to its rest energy is called a relativistic particle. The velocity of a relativistic particle is an appreciable fraction of the velocity of light. While the total energy (Equation (1.1) or (1.2)) of the moving particle depends on its motion relative to its surroundings, the rest energy (Equation (1.3)) is a characteristic independent of motion. Equation (1.3) expresses the equivalence of energy and mass. When a system of particles (a body) has lost energy ΔE, then according to Equation (1.3) its mass has decreased by $\Delta m = \Delta E / c^2$. The process of squeezing out energy from matter by mass decrease is of interest not only for astronomers but for any inhabitant of the Earth. The chemical processes (such as burning of fossil fuels) are very inefficient, because they lead to a relative decrease $\Delta m/m$ of only 10^{-10} or so. Nuclear reactions (such as those in stars) are more efficient because the relative mass decrease is of the order of 10^{-3}. By a gravitational contraction of a star even a few tenths of its rest energy may be released. However, the most efficient process is annihilation of matter and antimatter with complete conversion of the rest energy into radiation. Particles associated with radiation, i.e. photons, have zero rest mass.

1.1.3. MOMENTUM

The momentum of a particle with mass m_0 and velocity v is

$$p = \frac{m_0 v}{\sqrt{1 - \frac{v^2}{c^2}}} \tag{1.4}$$

which for small velocities $v \ll c$ reduces to the classical expression $p = m_0 v$. A relation between energy and momentum

$$\frac{E^2}{c^2} = p^2 + m_0^2 c^2 \tag{1.5}$$

follows from Equations (1.1) and (1.4). For zero-mass particles

$$p = \frac{E}{c}. \tag{1.6}$$

Conservation of momentum and energy applied to the decay of particles has allowed determination of their mass.

1.1.4. SPIN

Particles have their own angular momentum called spin. It is an intrinsic property and cannot be changed. We visualize the spin as gyroscope rotation. Conservation of angular momentum in decay or production reactions permits the determination of the spin of involved particles. Its values may be 0, $1\,\hbar$, $2\,\hbar$ for bosons or $\frac{1}{2}\hbar$, $\frac{3}{2}\hbar$ for fermions. The unit of spin is $\hbar = h/2\pi = 1.05 \times 10^{-27}\ \mathrm{g\,cm^2\,s^{-1}}$. For a particle with spin j its angular momentum around some axis is restricted to integrally separated numbers from $-j$ to $+j$. For example a particle with spin $j = \frac{3}{2}$ may be found in a magnetic field with angular momentum $\frac{3}{2}$ or $\frac{1}{2}$ or $-\frac{1}{2}$ or $-\frac{3}{2}$ and no other value is possible.

Particles are divided in two groups according to spin: fermions with half-integer spin (at the left in Figure 1.3) and bosons with integer spin (right part of the figure). The behaviour of fermions is governed by the Pauli principle which leads among other things to degeneration of very dense matter, e.g. in white dwarfs.

1.1.5. ELECTRIC CHARGE

Electric charge is an important characteristic of elementary particles, but its true nature is not yet known. Some particles bear a positive charge (e.g. proton), others are negatively charged (e.g. electron) and the rest have no charge (e.g., neutron, neutrino, photon). The quantity of charge -1.6×10^{-19} coulomb – is the smallest, indivisible and natural unit of electricity, attributed with incredible precision to all charged particles irrespective of the other properties like mass, spin, etc. If a difference should be found, then it must be less than 10^{-17} of the electron charge.

Electric charge is the source of electric forces in the surrounding space i.e. of the electrostatic field. When in motion it gives rise to magnetic forces (magnetic field). If accelerated it emits photons, quanta of the electromagnetic field.

Electric charge Q is conserved in all processes ($\Delta Q = 0$). It cannot be destroyed nor created. This explains why an electron left by itself cannot decay into a lighter particle: there exists none with electric charge.

1.1.6. BARYONIC NUMBER

The heavy particles (in the upper left corner of Figure 1.3) are called baryons. Left to themselves the baryons decay. The only stable baryon is the proton. If it decays at all then its lifetime should be longer than 10^{22} yr as shown by experiments. Such processes as

$$(p \rightarrow \mu^+ + \gamma \quad \text{or} \quad p \rightarrow e^+ + \gamma) \tag{1.7}$$

have *never* been observed, though they are energetically possible and the electric charge would be conserved. In all observed processes the number of baryons is always conserved. To describe the baryon conservation the baryonic number N has been introduced:

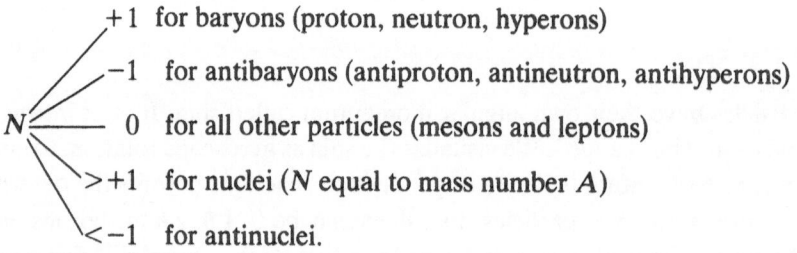

N
+1 for baryons (proton, neutron, hyperons)
−1   for antibaryons (antiproton, antineutron, antihyperons)
0   for all other particles (mesons and leptons)
>+1   for nuclei (N equal to mass number A)
<−1   for antinuclei.

The conservation law is then

$$\Delta N = 0 \tag{1.8}$$

which means that the sum of baryonic numbers is not changed by any process. The conservation law (Equation (1.8)) explains the stability of protons: there exists no lighter baryon than the proton. Experiments (with a very large number of protons) have shown that if the proton decays – its lifetime must be longer than 10^{22} yr, that is 10^{12} times longer than the age of the universe.

1.1.7. LEPTONIC NUMBER

The leptonic number has been introduced for similar reasons. Leptons are light fermions and leptonic number l is defined

$$l \begin{cases} +1 & \text{for leptons (electron, muon, neutrino)} \\ 0 & \text{for non-leptons (baryons and bosons)} \\ -1 & \text{for antileptons (positron, positive muon, antineutrino)} \end{cases}$$

The conservation law for leptonic number is then expressed as

$$\Delta l = 0. \tag{1.9}$$

The sum of all the l before a reaction equals the sum of l's afterwards. For example, materialization of high energy photons γ

$$\gamma \rightarrow e^- + e^+ \tag{1.10}$$

conserves the leptonic number, because l of an electron (e^-) is plus one and of a positron (e^+) is minus one.

Experiments with Na I give the lower limit for lifetime of electrons: if they decay at all, then after more than 10^{17} yr.

1.1.8. ISOSPIN

The strong interaction of nucleons in a nucleus does not depened on electric charge. Proton–proton, neutron–neutron and neutron–proton interactions are all alike. There is not much difference between neutral and charged nucleons. The difference is expressed by the isospin.

Figure 1.3 shows that particles may be grouped into multiplets, such as p and n; Σ^+, Σ^0 and Σ^-; π^+, π^0 and π^-. Particles of the same multiplet have the same spin, the same baryonic number and approximately the same mass, but their electric charge is different. The multiplet can be considered as different states of the same particle. By similarity with optical multiplets a new quantum number I (isospin, isotopic spin, isobaric spin) for particle multiplets is introduced.

To determine isospin e.g., for nucleons we move the charge origin in Figure 1.3 to the middle, viz. to $+e/2$. Then a new charge is assigned to the neutron and proton, viz. $-e/2$ and $+e/2$, respectively, just as the Pauli spin has two projections $-\hbar/2$ and $+\hbar/2$. A multiplet with isotopic spin I has $(2I+1)$ different charge states. The nucleonic doublet with isospin $I = \frac{1}{2}$ is in fact one particle called a nucleon, which exists in two charge states viz. as proton and neutron. The proton has isospin state $+\frac{1}{2}$, while the neutron has isospin state $-\frac{1}{2}$. Another example is the pion, which may exist in three different charge states (pi plus, pi zero, pi minus) and therefore $I = 1$. There is of course a fundamental difference between spin, which is a vector in ordinary space, and isospin, which is a vector in an artificial three-dimensional charge space. Spin characterizes angular momentum while isospin corresponds to electric charge.

1.1.9. STRANGENESS

Strangeness has been introduced to explain a strange behavior of hyperons and K mesons. These particles are produced by strong interactions and decay by weak interactions. The hyperons were always observed to be created with K mesons (K^+ or K^0 but never K^-). The creation in collisions of hadrons is very fast, which is characteristic for strong interactions. For example:

$$\pi^+ + p \rightarrow \Sigma^+ + K^+$$
$$\pi^- + p \rightarrow \Lambda^0 + K^0 \tag{1.11}$$
$$\pi^+ + p \rightarrow \Xi^0 + K^+ + K^+.$$

Strangeness cannot be measured directly. It may be found for the multiplets in Figure 1.3. The charge center of the multiplet is shifted by a certain amount with respect to the charge center of the nucleon doublet or pion triplet. The strangeness S is then equal to the double shift. Thus for Λ^0 the charge shift is $-\frac{1}{2}$ so that its $S = -1$; also for Σ hyperons $S = -1$; for Ξ doublet the shift is -1 and $S = -2$ while the strangeness of the proton and neutron is zero. For K^+K^0 doublet the shift with respect to the pion triplet is $+\frac{1}{2}$ so that the strangeness of kaons is $+1$, while for pions $S = 0$.

Particles with strangeness different from zero are called strange particles. They are created in a very short time ($\sim 10^{-23}$ s) and their strangeness is conserved

$$\Delta S = 0. \tag{1.12}$$

This may be seen in reactions (1.11), for example

with S
$$\left.\begin{array}{c} \pi^+ + p \rightarrow \Xi^0 + K^+ + K^+ \\ 0 + 0 = -2 + 1 + 1. \end{array}\right\} \tag{1.11a}$$

The reaction is allowed because strangeness is conserved. On the other hand

with S
$$\left.\begin{array}{c} \pi^+ + p \rightarrow \Sigma^+ + \pi^+ \\ 0 + 0 \neq -1 + 0 \end{array}\right\}$$

is forbidden because Equation (1.12) does not hold.

Decay of strange particles violates Equation (1.12) and is therefore very slow, with decay time $\geq 10^{-10}$ s, which is typical for weak interactions. To be specific:

with S
$$\left.\begin{array}{c} \Sigma^+ \rightarrow p + \pi^0 \\ -1 \neq 0 + 0 \end{array}\right\} \tag{1.13}$$

where strangeness is not conserved. The decay occurs via weak interactions.

1.1.10. PARITY

Parity is another property of elementary particles which corresponds to mirror reflection of space coordinates. It is a symmetry property of a wave function. It may be either plus or minus if the wave function is unchanged by reflection or changed (i.e., even or odd). Parity is conserved in strong and electro-magnetic interactions and violated in weak interactions.

1.2. Antiparticles

It follows from the principles of relativity and quantum mechanics that to each particle there should exist an antiparticle with the same mass and the same spin. The other quantum numbers (electric charge, isospin, strangeness, baryonic number, leptonic number) have the same magnitude as for normal particles but the sign is reversed. The relation of particles and antiparticles is seen in Figure 1.3. It should be stressed that it is quite arbitrary to call electrons, protons, and neutrons particles rather than antiparticles; it is however more natural to consider ourselves (and the environment we live in) as consisting of matter rather than of antimatter.

For reasons of clarity, the matter of which our environment consists is called koinomatter, as opposed to antimatter, which consists of antiparticles.

The laws of conservation are valid for reactions involving particles and antiparticles. For example:

$$p+p \rightarrow p+n+p+\bar{p}+\pi^+$$

$$\text{electric charge} \quad +1+1 = +1+0+1-1+1 \tag{1.14}$$

$$\text{baryonic number} + 1 + 1 = +1+1+1-1+0$$

where $\bar{p}$ denotes antiproton.

Two processes have to be mentioned when talking about antimatter: annihilation (Figure 1.1) and pair production. A particle and its antiparticle annihilate in mutual collision and their energy is converted into photons or mesons. For example:

$$e^- + e^+ \rightarrow 2\gamma \tag{1.15}$$

and

$$p + \bar{p} \rightarrow 2\gamma \tag{1.16}$$

with energy 0.5 MeV for each photon in Equation (1.15) and 938 MeV for each of the photons in Equation (1.16).

Pair production (materialization) is a reverse process to annihilation. It is the formation of a particle and its antiparticle from a photon with sufficiently high energy. The pair production occurs through electromagnetic interaction of the

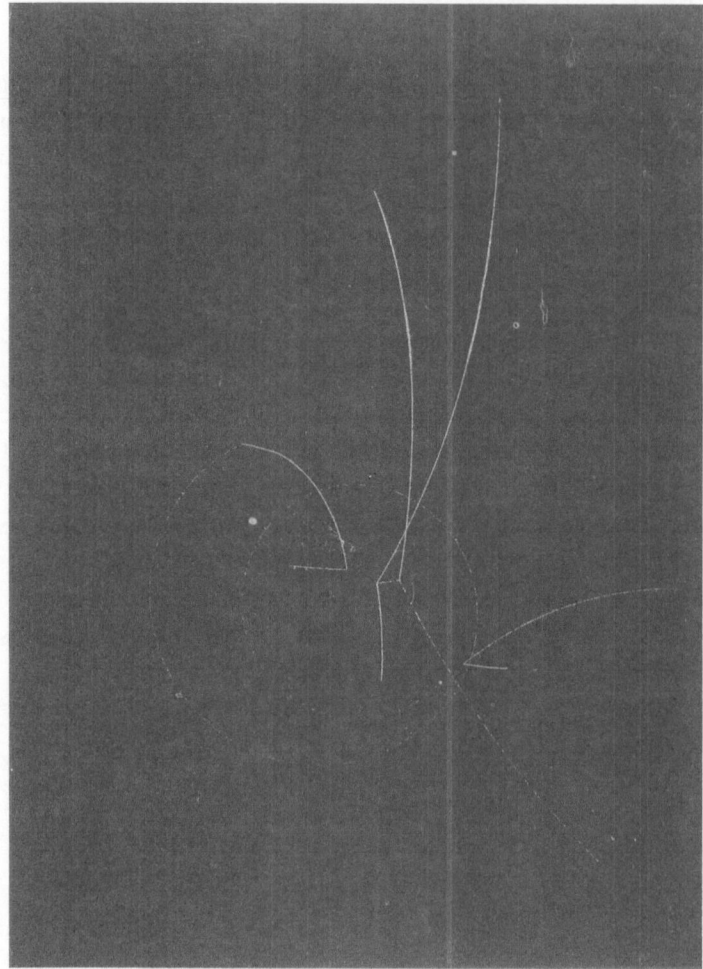

Figs. 1.1a–b.   An antiproton $\bar{p}$ enters from the top into a liquid hydrogen chamber (Saclay/École Polytechnique, Paris). The antiproton annihilates with a proton in point A. In the annihilation mesons K^0, K^- and π^+ as well as a π^0 are formed. The latter cannot be observed directly and is found by kinematic analysis of the event. The K^0 decays in B into π^+ and π^-, the π^+ decays in C, giving μ^+ and ν_μ (Equation (1.25)); the μ^+ decays at D into e^+, ν_e, $\bar{\nu}_\mu$ according to Equation (2.6). The K^- interacts in E, giving Λ^0 and π^0. The Λ^0 decays in F into a proton, which stops in G and a π^-. The π^0 is not seen since it is neutral and can leave no track. The π^+ produced in A scatters on a proton in H (CERN).

photon (or a high-energy particle) with the field of an atomic nucleus or other particle (Figure 1.2). Another way of pair creation is through the de-excitation of an excited nucleus (the so called internal pair production). The best known example of pair production is the creation of an electron–positron pair in the field of an atomic nucleus N:

$$\gamma + N \rightarrow N + e^- + e^+. \tag{1.17}$$

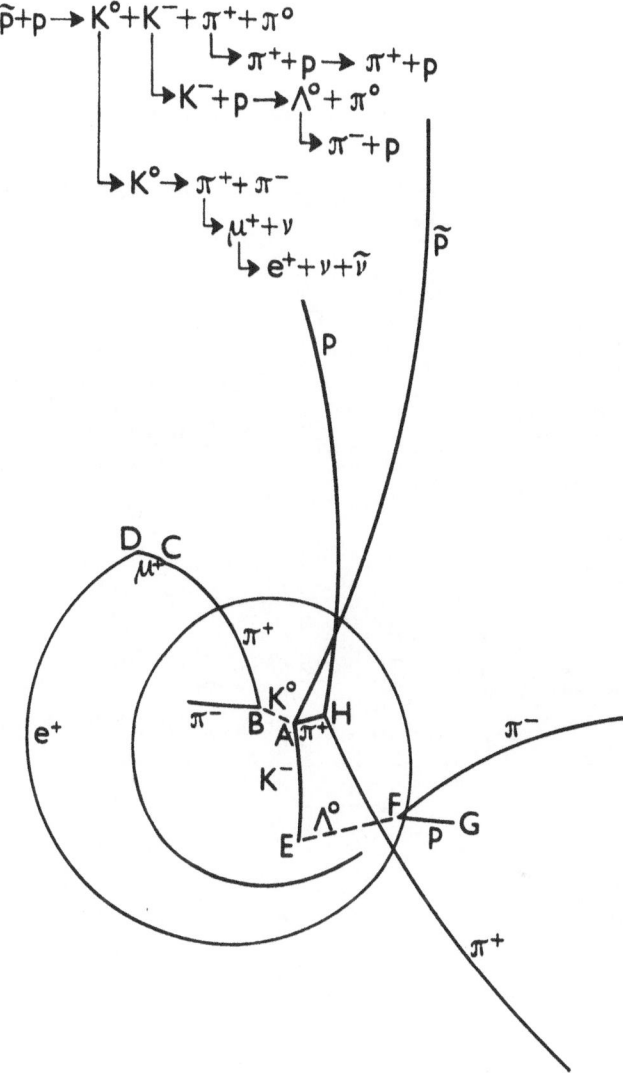

Fig. 1.1b.

The photon energy must exceed the combined rest energies of the produced particles i.e. 1 MeV. By analogy

$$\gamma + p \rightarrow n + \pi^+ \ (\gamma \text{ energy} > 140 \text{ MeV}) \tag{1.18}$$

$$\gamma + N \rightarrow N + p + \bar{p} \ (\gamma \text{ energy} \geqslant 2 \text{ GeV}) \tag{1.19}$$

$$\gamma + \gamma \rightarrow \nu + \tilde{\nu}. \tag{1.20}$$

Antiparticles are produced in large particle accelerators and generally in any place where high energy particles occur, e.g., cosmic ray particles in the terrestrial atmosphere, in interstellar space, in the Crab nebula, etc. But one still does not know

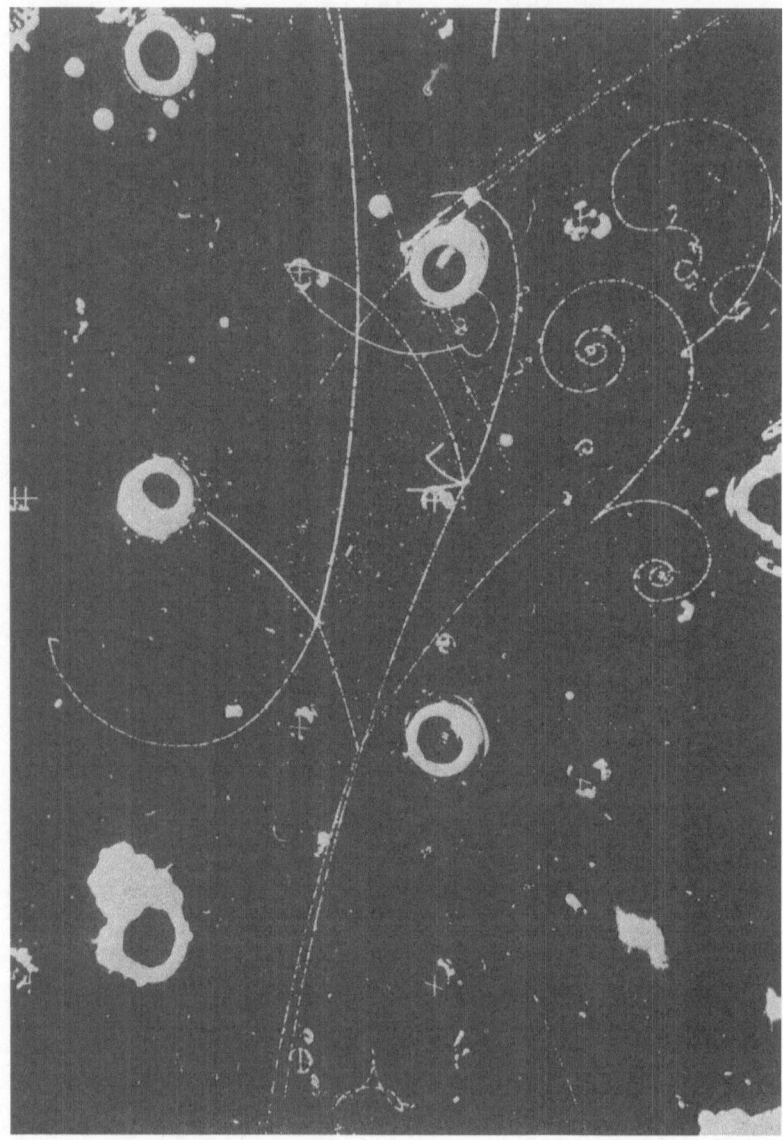

Figs. 1.2a–b.   Annihilation of an antiproton p̄ arriving from below (track C) with a proton of the liquid hydrogen (point A). Five particles are produced by the annihilation reaction: π^+, π^-, π^0 and two gamma photons. The gamma photons have enough energy for electron–positron pair formation in points D. Another process occurs in point B, where a particle in the bubble chamber at E, collides with a hydrogen nucleus (CERN).

$$\tilde{P}\, p \longrightarrow \pi^+ \pi^- \pi^\circ$$

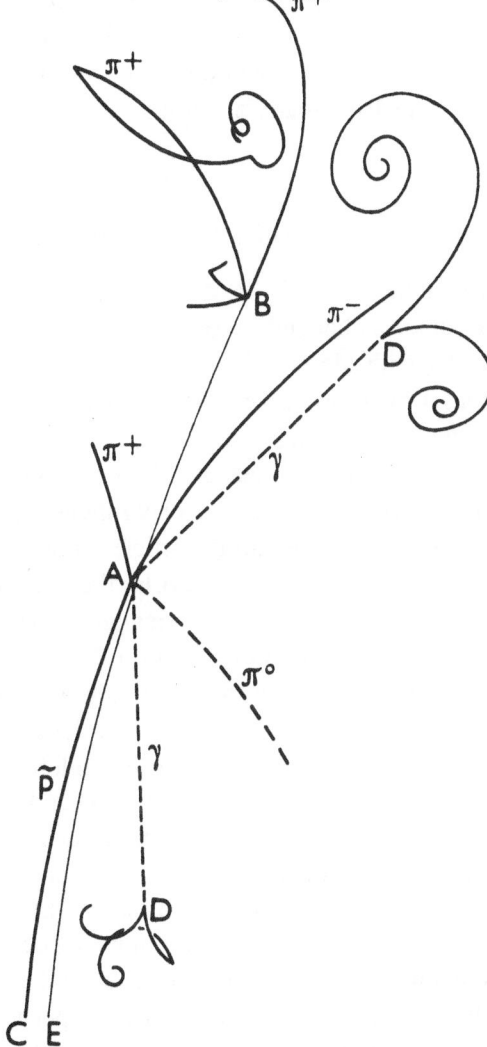

Fig. 1.2b.

whether antimatter exists in large quantities in our universe. Unfortunately a photon emitted by matter is the same as photon emitted by antimatter, so that by observing only electromagnetic radiation one cannot decide whether a star consists of normal matter (koinomatter) or antimatter. Neutrinos seem more promising in this respect: stars of koinomatter should be sources of neutrinos, while antistars are sources of antineutrinos. The sensitivity of present-day neutrino detectors is several orders of magnitude too low, however, to decide whether antimatter exists in large enough quantities to be important for the structure and evolution of the universe.

It is probable that many antiparticles existed in the very early phase of the history of the universe (see Chapter 5). But it is not known whether the number of particles was equal to the number of antiparticles or whether there was a slight asymmetry in favor of particles. The antimatter might have played an important role in the universe because its annihilation with koinomatter is the most effective mechanism for extracting the rest energy from material bodies.

1.3. System of Elementary Particles

The grouping of elementary particles according to their mass (or rest energy), spin, electric charge, baryonic number and leptonic number is represented in Figure 1.3.

Resonance particles with lifetimes of about 10^{-23} s are not listed. They decay strongly into hadrons, which may also be unstable. Their lifetime is so short that even if they moved with the velocity of light, they would cover only a distance comparable with the diameter of the atomic nucleus. They are not detected directly; one recognizes their existence from peaks in experimental plots representing cross section against energy of the bombarding particle. Though the study of resonance particles occupies many laboratories in the world; their nature is not yet completely understood. Are they entities similar to other elementary particles (e.g. excited states of baryons and mesons)? We could extend Figure 1.3 by adding resonances, but they are too many; it is not yet sure whether they are really elementary, and finally it does not seem that they are important in the structure of the universe and its evolution.

Baryons are heavy particles represented in the upper left corner of Figure 1.3. The group of baryons consists of two nucleons (proton and neutron) and hyperons $(\Lambda^0, \Sigma^+, \Sigma^0, \Sigma^-, \Xi^0, \Xi^-, \Omega^-)$. Nucleons, being constituents of atomic nuclei, are the most abundant baryons in the observable universe. Hyperons – heavier baryons – are unstable under normal conditions; they are produced by collisions of cosmic ray particles with atmospheric or interstellar atoms or by large accelerators (Figure 1.1). Due to their instability hyperons would be rare guests in the universe – if it were not for the highly condensed matter in degenerate stars (hyperon stars). There the Pauli principle forbids decay of hyperons, so that they are stable particles (see Part 3). But until now there is no estimate for the amount of hyperon matter in our Galaxy or elsewhere in the universe.

The baryons and baryon resonances appear to be different exited states of one particle only. In some very strong interactions they are really divested of all differences and appear as one baryon particle. The left upper part of Figure 1.3 resembles a diagram of energy levels from atomic spectroscopy. In a similar manner the four lepton particles (electron e^-, muon μ^-, neutrino ν_e and neutretto ν_μ) are sometimes considered to be different states of the same particle – the lepton.

Bosons are field quanta with integer spin (0, 1, 2 in units $\hbar = h/2\pi$, where h is the Planck constant 6.6×10^{-27} erg s). Each field is produced by a source (see Chapter 2) but it can propagate independently from the source, in the form of field quanta when

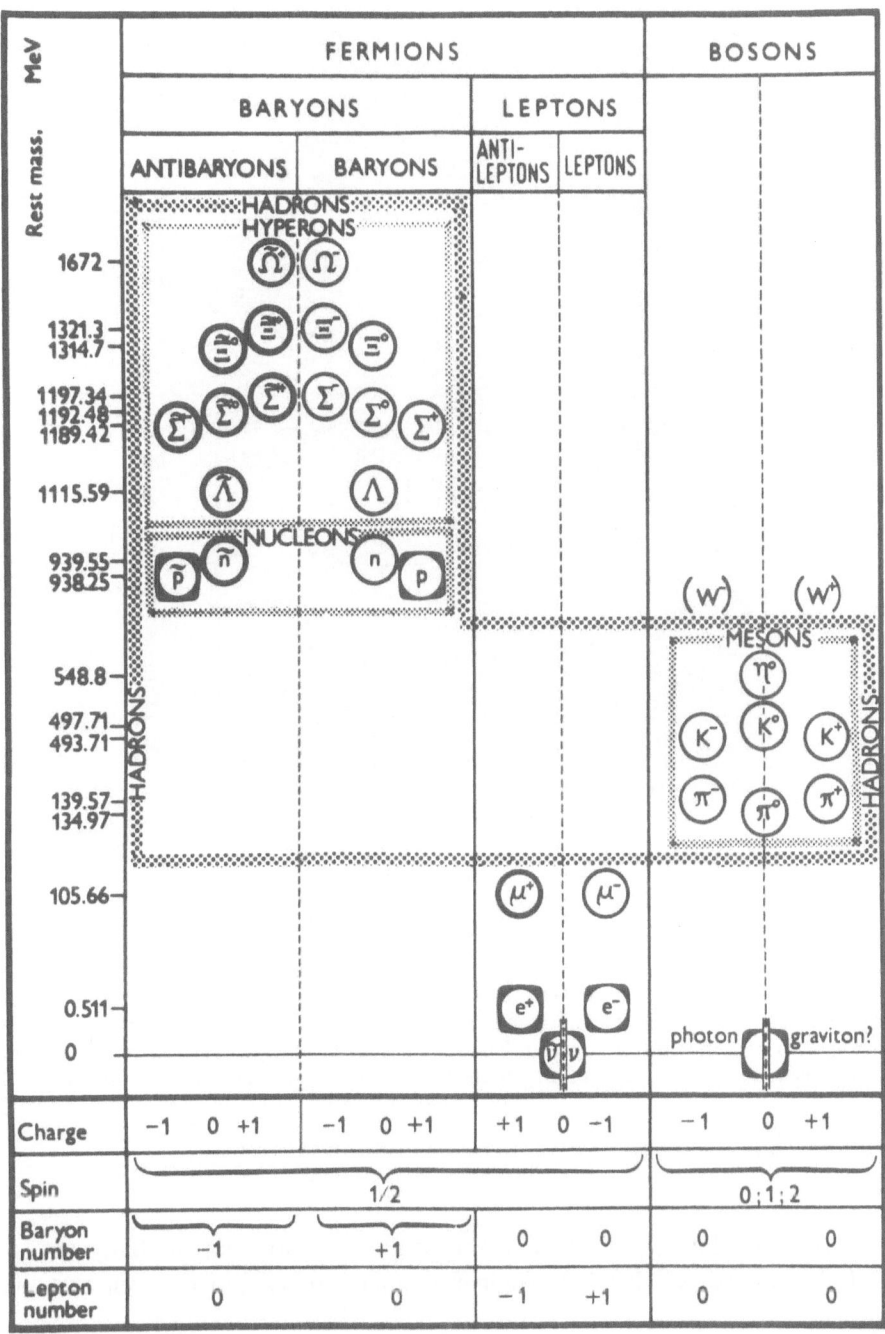

Fig. 1.3. Elementary particles classified according to their mass (leptons, mesons, baryons), spin (bosons and fermions), electric charge, baryon number and lepton number. Mesons and baryons are strongly interacting particles (hadrons). Baryons are subdivided into stable nucleons and unstable hyperons. No resonances are represented. The intermediate boson (W$^+$ and W$^-$) and the graviton are hypothetical particles.

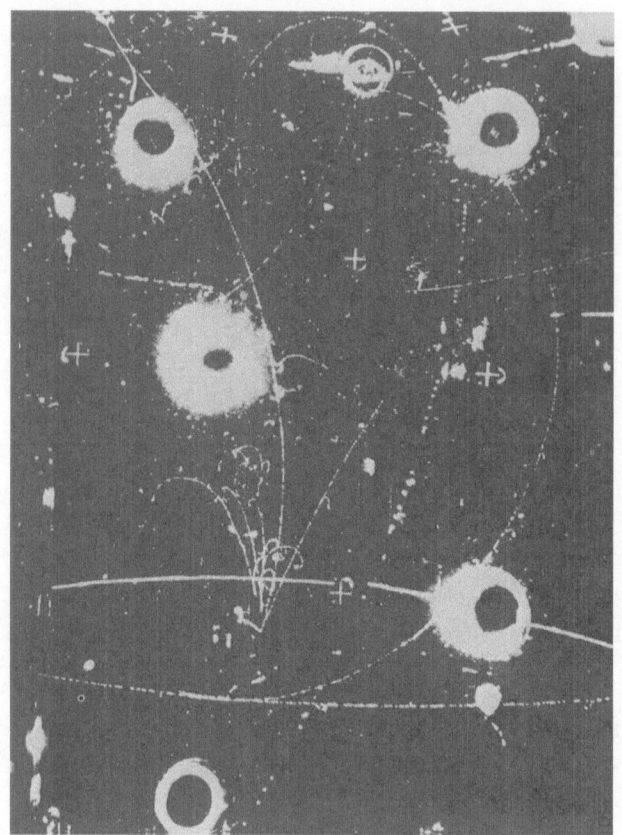

Figs. 1.4a–b.   This photograph shows a reaction in which a neutretto (A) strikes at (B) a nucleon of the chamber liquid, producing several gamma photons (C). The photons in turn produce electron–positron pairs (D) and above all a positive pion (H) and a negative pion (E). White rings (F) are due to the flashes which light up the inside of the chamber. The curvature of the charged particles in this and other photographs is caused by the Lorentz force; the bubble chamber is placed into a magnetic field (CERN).

the source is accelerated. The field quanta sometimes carry electric charge (π^+, π^-, K^+, K^-), strangeness (K^+, K^0), angular momentum, isospin or even rest mass (mesons).

Gravitons are hypothetical particles with a spin 2, which have not yet been observed, due to their very weak effects.

Photons are quanta of electromagnetic fields. They are produced by accelerated charged particles, above all by the lightest ones – electrons. For example:

$$e^- \rightarrow e^- + h\nu. \tag{1.21}$$

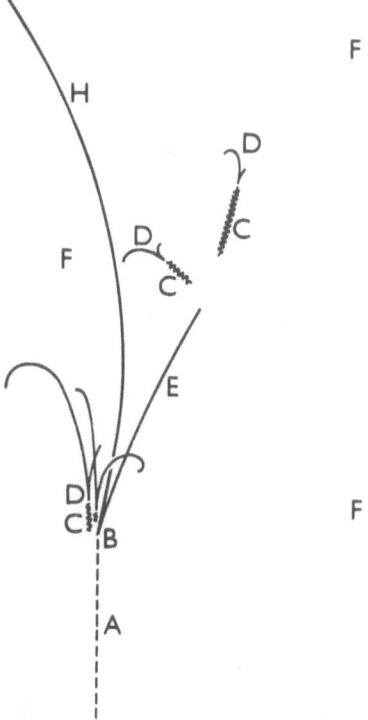

Fig. 1.4b.

Two particles in our table viz. Σ^0 and π^0 decay spontaneously into photons:

$$\Sigma^0 \to \Lambda^0 + \gamma \tag{1.22}$$

$$\pi^0 \to 2\gamma. \tag{1.23}$$

Photons have zero mass, spin 1 and always move with the highest velocity $c = 3 \times 10^{10}$ cm s^{-1}.

Pions (π mesons) and kaons (K mesons) are quanta of nuclear fields. They have no spin but they carry mass and charge, in contrast to photons and gravitons. Their source are baryons. They are emitted when a baryon is strongly accelerated or by a decay of hyperons, as for example (Figure 1.5):

$$
\begin{aligned}
\Lambda^0 &\to p + \pi^- \\
\Lambda^0 &\to n + \pi^0 \\
\Xi^0 &\to \Lambda^0 + \pi^0 \\
\Omega^- &\to \Lambda^0 + K^-.
\end{aligned}
\tag{1.24}
$$

The mesons themselves decay into leptons and γ quanta:

$$\pi^+ \rightarrow \mu^+ + \nu_\mu$$

$$\pi^- \rightarrow \mu^- + \tilde{\nu}_\mu$$

$$\pi^0 \rightarrow 2\gamma \tag{1.25}$$

$$K^0 \rightarrow \pi^+ + \pi^-$$

$$K^0 \rightarrow \pi^0 + \pi^+ + \pi^- \text{ etc.}$$

There are other different ways of meson decay.

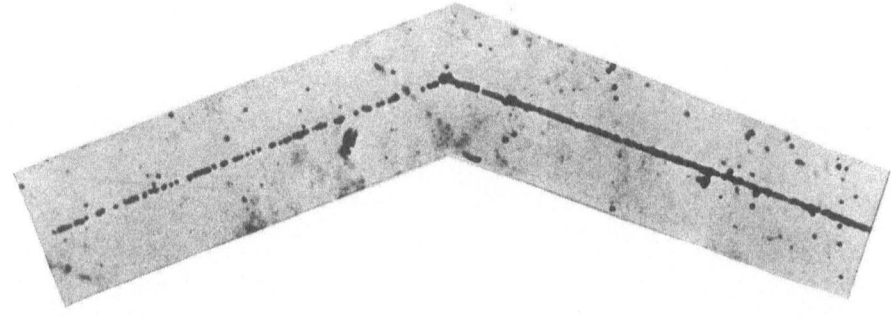

Fig. 1.5.   Decay of a Λ-hyperon into a proton (the full track to the right) and a π^--meson (interrupted track to the left). The charged particles ionize atoms of grains in a special (nuclear) emulsion and leave a latent image of the particle track (Decay Equation 1.24).

For the sake of completeness one should mention the heavy boson or W-boson – a hypothetical field particle for weak interactions. Its mass should be larger than the mass of the K meson, otherwise the heavy boson would be produced by the decay of the kaons.

A detailed discussion of field quanta and forces in the universe will be found in Chapter 2.

Summary

Elementary particles are fundamental units for the structure of the universe. According to their rest mass they are classified as baryons (heavy particles), leptons (light particles) and mesons (medium weight particles). Mesons belong to a group of field particles, bosons, which are produced by baryons and leptons, either by acceleration or decay.

The particles important for the structure of the universe are ordered in Figure 1.3. Short-lived resonances are not included. All the resonances and many of the particles listed in the table are unstable – they decay into lighter and stable particles. While the number of unstable particles is still growing through new discoveries, the list of the few stable particles does not change.

Most and possibly all elementary particles may be created by materialization of energy. The minimum amount of energy E required for production of a group of particles with total mass m follows from the equivalence equation $E = mc^2$.

Under high densities unstable particles (such as neutrons, hyperons, mesons) acquire stability. On the other hand stable particles (e.g., electrons, protons) may be destroyed in mutual collisions with their antiparticles. Nothing in the material universe has a secure existence, if the stable fundamental units do not have a secure existence.

PARTICLES AND FORCES

No particle in our universe is completely isolated from the influence of other particles. They exert forces on each other i.e., they interact by means of their fields.

Each field is produced by a source in the particle itself. If the source is accelerated the field propagates in the form of field quanta independently from the source. All the field quanta are bosons (photon, graviton, meson, heavy boson). There are four known types of fields and quanta (Table 2.1). The four types of forces (i.e., interactions of elementary particles) are of fundamental importance for the structure and evolution of the universe.

TABLE 2.1

Properties of interactions

Source	Interaction	Field quantum	Relative strength	Range cm	Example
baryon, meson	strong	pion, kaon	1	10^{-13}	nuclear forces
electric charge	electro-magnetic	photon	10^{-2}	∞	atomic and mole-cular forces
lepton, meson baryon	weak	(W-boson)	10^{-14}	$<10^{-15}$	beta decay
mass of any particle	gravitation	(graviton)	10^{-40}	∞	large-scale cosmic attraction

Considering all the particles in Figure 1.3 one can see that each of them is endowed with one or more fields. One can roughly say: the weaker the interaction the more particles participate in it. Baryons and mesons experience strong interactions, all charged and some neutral (except neutrinos and gravitons) are governed by electromagnetic interactions, all but gravitons experience weak interactions and all the particles without exception (including zero mass particles) experience gravitational interactions. This means that a particle experiencing an interaction experiences automatically all the weaker interactions.

The hierarchy of the four interactions is connected with conservation of the characteristic quantities of elementary particles. One may see in Table 2.2 that all conservation laws are valid for strong interactions. Strong interactions are therefore

TABLE 2.2

Conservation laws are either valid $(+)$ or violated $(-)$

Quantity / Interaction	Energy	Momentum	Moment of momentum	Charges: electric, leptonic, baryonic	Strange-ness	Parity	Isospin
strong	+	+	+	+	+	+	+
electromagnetic	+	+	+	+	+	+	−
weak	+	+	+	+	−	−	−

allowed by all the laws and hence are very fast. On the other hand strangeness, parity and isospin are not conserved in weak interactions. Therefore the weak interactions trespass against three conservation laws and are nearly forbidden; hence their sluggishness. Gravitation is omitted from Table 2.2 because its role in elementary processes is unknown.

This classification of interactions has been used since 1948. It is very important for understanding the nature of elementary particles themselves as well as for understanding the structure and evolution of the universe. In every structural feature or evolutionary process one or more interactions are involved.

2.1. Nuclear Forces

Nuclear forces (strong interactions) act everywhere in the universe where two hadrons are (or collide) at a distance of about 10^{-13} cm: in the nuclei of atoms, in the cores of stars, in interstellar space, in meteorites, on the surface of the Moon and other satellites, in the atmosphere of planets, etc. The close approach of two hadrons may be hampered by their electric repulsion, so that a rather high energy (either kinetic or pressure) is necessary to overcome the repulsion. The electrostatic repulsion between two protons in a nucleus is equal to the gravitational pull of the Earth on a mass of 25 kg at the Earth's surface. The nuclear attractive forces between the protons outweighs the electric repulsive force at a distance of 10^{-13} cm. When the electromagnetic interactions are allowed for, it is found that the interaction between proton and proton is exactly the same as between proton and neutron or neutron and neutron. The strong interactions are independent of the electric charge of the hadrons.

2.1.1. ATOMIC NUCLEUS

The atomic nucleus is the central part of the atom which occupies about 10^{-14} of its volume but represents almost the whole atomic mass. The atomic nucleus is a system of nucleons held together by the exchange of quanta of strong interaction – pions. Over 1400 nuclides are known (nuclide is a species of nucleus characterized by atomic number and mass number). The number of protons in a nuclide ranges from 1 to 104 and is called the atomic number Z. The number of neutrons varies from zero to 157. The total number of nucleons in a nuclide is called mass number A and it determines its mass. For light nuclei A is usually close to $2Z$, but for heavier elements the neutron number $(A - Z)$ is larger than Z.

More than 280 stable nuclides are known. A measure of stability of a nucleus is its binding energy. It is the energy required to decompose it into free nucleons. It may also mean the energy necessary to remove a nucleon from the nucleus (in the latter case it is also called separation energy). The mass of a nucleus $m(Z, A)$ is smaller than the sum of the masses of individual nucleons outside the nucleus:

$$\Delta m = [Zm_p + (A - Z)m_n] - m(Z, A). \tag{2.1}$$

The difference between the sum of the masses of the constituent nucleons and the mass of the nucleus is called the mass defect. The binding energy of the nucleus is then

$$E_b = (\Delta m)c^2 \tag{2.2}$$

and the binding energy for one nucleon is E_b/A. For example: the nucleus of He (α-particle) has a binding energy 28.3 MeV and a specific binding energy per one nucleon 7.07 MeV. For the majority of nuclei the binding energy per one nucleon is about 8 MeV. The binding energy per nucleon depends on the mass number of the nucleus, as may be seen in Figure 2.1. It is largest for the group of elements around Fe. It is smaller for light nuclei because the ratio of surface to volume for such nuclei is large and the surface nucleons are less strongly bound than nucleons in the interior. The nuclei heavier than Fe are also less strongly bound because electrostatic repulsion increases with the number of protons. In the group of elements near Fe both tendencies are balanced, hence their stability. The shape of the curve in Figure 2.1 indicates that energy may be liberated either by the synthesis of light nuclei (e.g. in thermonuclear reactions in stars) or by fission of heavy nuclei (as in nuclear reactors).

2.1.2. CHART OF NUCLIDES

A graphical representation of the relationship between the nuclides is often used (Figure 2.2). It is called either a chart of nuclides or an isotope chart. It is a simple grid of points in cartesian coordinates: the integer numbers on the horizontal axis correspond to the number of neutrons in a nuclide, while the vertical axis represents the number of protons. For example, the heaviest stable nuclide Bi ($Z = 83$, $A = 209$) is represented by a point (126, 83) and the lightest nuclide H (1, 1) by the point (0, 1). A chart of nuclides gives a visual representation of the structure of nuclides and nuclear processes (see Section 5.4).

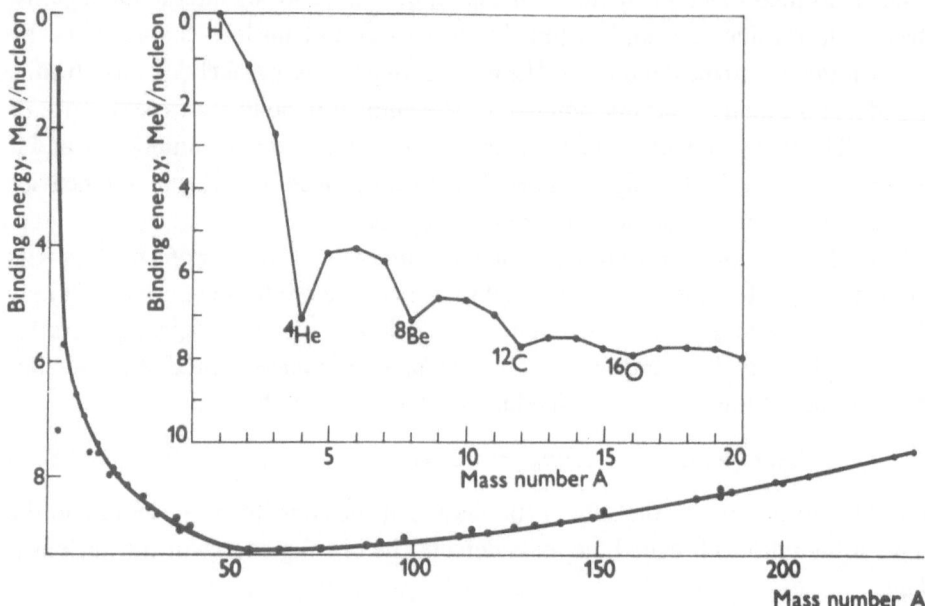

Fig. 2.1. Binding energy per one nucleon. At the bottom of the curve are nuclei of the iron group
($50 \leqslant A \leqslant 65$). The nuclei of small mass number (i.e. on the left slope of the curve) are characterized by a
large surface-to-volume ratio. Their fusion into heavier nuclei (Table 2.3 and Equations (5.20), (5.21),
(5.23), (5.24) etc.) increases the binding energy per one nucleon. It represents the main source of
luminosity of stars during their thermonuclear evolution phase (Sections 5.4.2, 5.4.3). In heavy nuclei
electrostatic repulsion between protons becomes important and energy may be produced by nuclear
fission.

One could represent the binding energy E_b of each nuclide as a third coordinate
perpendicular to the plane of the chart of the nuclides – let us say below the plane.
The resulting surface represents a valley of stability. The curve in Figure 2.1.
corresponds roughly to the cut along the bottom of the valley. On its left slope are
proton-rich nuclides and on the right slope neutron-rich nuclides. The nuclides far
from the bottom are unstable, with a tendency to decay and move towards the
bottom.

2.1.3. THERMONUCLEAR REACTIONS

By fusing light nuclei into a heavier one, the difference of binding energies may be
liberated. The process of fusion corresponds to the shift to the right in Figure 2.1. The
two colliding nuclei must have sufficiently high kinetic energies to overcome their
mutual electric repulsion. The kinetic energy may be due to high temperature, as in
the case of stellar interiors; or the kinetic energy may be non-thermal, as in collisions
of cosmic-ray particles or fluxes of high energy particles in stellar atmospheres. The
nuclear reactions under high temperatures are called thermonuclear reactions.

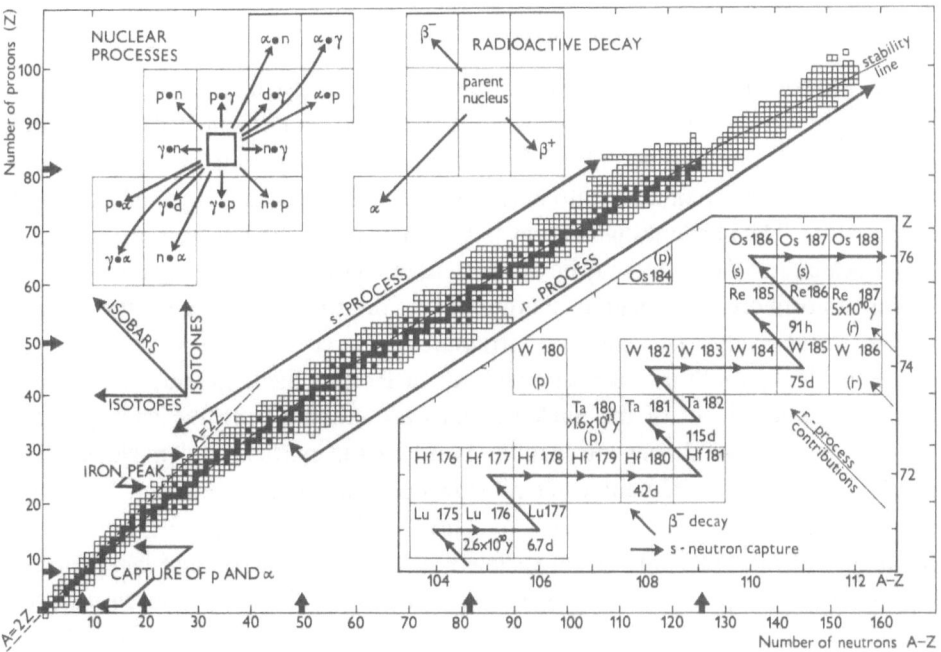

Fig. 2.2. Chart of nuclides (isotope chart) represents structure of different species of atomic nuclei. The dark squares represent stable nuclides and they lie along the stability line – at the bottom of stability valley. The unstable nuclides have a tendency to get towards the stability valley (by β^- decay from the right slope, by β^+ decay or electron capture from the left slope). Main thermonuclear transformations (nuclear processes and radioactive decay) are represented at the top. Important thermonuclear processes responsible for production of the nuclides are indicated by arrows along the valley of stability. One nuclide may be, however, produced by several processes and the arrows serve only for a rough orientation. An example of the s-process is in the lower right corner.

The main phases of stellar evolution are characterized by thermonuclear fusion of protons into He nuclei, of He nuclei to C nuclei etc., as described in detail in Section 5.4. A massive star, for example, with a mass of 10 solar masses, emits about 10^4 times more energy than our Sun. The star produces such huge energy fluxes by thermonuclear reactions deep in the stellar interior, gradually transforming H into He, and further heavy elements up to Fe, liberating at the same time the binding energy shown in Figure 2.1.

Thermonuclear reactions in the stellar interior have a double effect: first, heavier atoms have been and are being produced from primeval H (nucleosynthesis); second, the liberated binding energy is transformed into the radiation of stars. The life times of stars and the existence of all the chemical elements (with the exception of the lightest ones) are therefore a consequence of strong interactions between nucleons in the deep interior of stars. The dependence of the structure and evolution of stars and also the formation of heavier elements on thermonuclear reactions are the subject of Section 5.4.

2.1.4. NUCLEAR INTERACTIONS WITH COSMIC RAYS

In order to become a part of the nucleus a nucleon must give up a fraction (about 0.1 to 0.8%) of its rest energy. And conversely, to escape from the nucleus the nucleon must get back the binding energy from some high energy particle. High energy particles which may penetrate into nuclei are abundant in cosmic rays and terrestrial accelerators. The incident high energy particle is most often a proton, α-particle or light nucleus and the target may be the nucleus of an interstellar atom, a nucleus in a stellar atmosphere, in a planetary atmosphere, on a meteorite or on the Moon's surface or the nucleus of a substance irradiated by a particle accelerator.

The incident particle interacts directly with nucleons of the target nucleus. Some of them are knocked out of the nucleus. Pions are created from the energy of the incident particle by strong interactions between the particle and nucleons. The residual nucleus is excited and can emit slow nucleons (p, n) and nuclear fragments (e.g. ^{2}H, ^{3}He, Li, Be, B). The nuclear reaction in which several particles or nuclear fragments are ejected from a target nucleus is called spallation (spallation reaction or high energy nuclear reaction) – from the verb 'to spall' meaning to chip with a hammer.

If the target nucleus is heavy ($A > 80$), then the residual nucleus may de-excite by splitting into two roughly equal parts. The process is called nuclear fission and is accompanied by the emission of neutrons and γ-photons. As the elements with $A > 80$ are rather rare, fission reactions seldom occur in nature.

The spallation reactions of major importance in the universe are those involving abundant nuclei such as ^{4}He, ^{12}C, ^{14}N, ^{16}O and leading to products ^{2}H, ^{3}H, ^{3}He, Li, Be, B, which are relatively rare in the universe. As for spallation in the interstellar space, there are two possibilities: (a) the abundant nuclei He, C, N, O in interstellar gas are bombarded by cosmic rays. Then the spallation products also become part of the interstellar gas. This process is the main source of D, Li, Be and B in the universe. (b) on the other hand if the abundant nuclei He, C, N, O are the high energy cosmic ray particles incident on interstellar target nuclei, then their spallation products ^{2}H, ^{3}H, ^{3}He, Li, Be, B become also cosmic-ray nuclei. Measurements show that the latter nuclei are about a million times more abundant in cosmic rays than in stellar material. Accelerator experiments indicate that cosmic rays have to traverse about $3 \, \mathrm{g \, cm^{-2}}$ of interstellar gas. As the density of interstellar gas is about $3 \times 10^{-24} \, \mathrm{g \, cm^{-3}}$, the cosmic rays had to traverse a distance $3 \, \mathrm{g \, cm^{-2}}/3 \times 10^{-24} \, \mathrm{g \, cm^{-3}}$ which is 10^{24} cm or about 10^6 light years. As the cosmic rays move with velocities only slightly less than the velocity of light, the average of a cosmic ray particle is about one million years.

Strong interactions in the terrestrial atmosphere act in a double manner: as a glue in stable nuclei (mainly nitrogen and oxygen nuclei) and as high energy reactions of cosmic rays with those nuclei. The products of the latter reactions (nucleons, hyperons, pions, kaons) induce further spallation reactions in lower atmospheric layers, etc. This chain of spallation reactions is called nuclear cascade or nucleonic cascade. Propagating to lower layers, the host of secondary particles (nucleons, mesons, nuclear fragments) proliferate by strong interactions while other secondaries (electrons, photons, muons and neutrinos) participate in electromagnetic and weak interactions. In a low density medium like the Earth's atmosphere many pions decay into muons. A fraction of the muons decay while still in the atmosphere, but many of them react with the Earth's surface and penetrate to depths of hundreds of meters. One primary cosmic-ray particle of 10 GeV generates only an unimportant cascade with a small number of secondary products. Such a small cascade dies out after a few generations, high in the atmosphere.

A primary particle with energy $>10^{17}$ eV generates a huge nuclear cascade (called atmospheric shower) with a large number of generations of secondary particles. The number of secondaries reaching the Earth's surface may be many millions in one cascade. The whole atmospheric shower with many strong, electromagnetic, and weak interactions proceeds so fast through the atmosphere that its overall duration is less than 0.001 s. Most particles fly in the direction of the primary particle and are concentrated in a disk of about 1 m thickness and a few meters diameter (core of the shower), when it arrives at the Earth's surface. The particle density decreases with distance from the core and may be traced at a distance of a few hundred meters. From the measurements of a shower at the Earth's surface the energy of a primary parent particle may be determined. The highest energies measured in this way are about 10^{20} eV per particle. One does not understand what process in the universe can concentrate such a fantastic energy (about 10 J) in such a tiny particle as the proton.

All the bodies of the solar system are exposed to bombardment by primary cosmic particles. Their interaction with the surfaces of meteorites, the Moon, satellites, asteroids and interplanetary dust induces spallation reactions in atomic nuclei of these solid cosmic bodies. As a result, a series of new nuclei (called cosmogenic nuclei) remains in the solid bodies as a trace of the high energy nuclear reactions. Some of the cosmogenic nuclei are stable, others are radioactive. Helium, Ne, Ar, and Xe belong to the stable products. Among radioactive nuclei we find ^{10}Be (half-life 2.7×10^6 yr), ^{32}Si (700 yr), ^{36}Cl (3×10^5 yr), ^{14}C (5770 yr), ^{55}F (3 yr), ^{57}Co (270 days) and others. The amount of the cosmogenic elements depends upon the duration of exposure to cosmic radiation. This duration – known as 'exposure age' – is an important parameter for the evolution of the whole solar system in general and of meteorites in particular. Prior to falling on the Earth, meteorites are exposed to cosmic ray bombardment as they move in the interplanetary space. The exposure age of meteorites is much shorter than their proper age – called 'solidification age' – according to many measurements. These results are consistent with the idea that the meteoritic material had been at first part of a large parent body (an asteroid), shielded from cosmic radiation. After a cataclysmic collision the meteorites broke loose from the parent asteroid and have been exposed to cosmic radiation since that moment.

Cosmogenic elements in meteorites and lunar rocks are also a source of information regarding the cosmic ray flux in past times. The cosmic ray flux is an important indicator of stellar activity in our Galaxy, above all in the neighborhood of the Sun. From meteoritic research one concludes that no catastrophic event affected any of our nearest stellar neighbors during the past 10^8 yr.

2.1.5. STRONG INTERACTIONS IN STELLAR ATMOSPHERES

We have discussed high energy nuclear reactions induced by cosmic rays in interstellar space, in the terrestrial atmosphere and in the solid bodies of the solar system. In stellar atmospheres protons and other nuclei are accelerated by magnetic fields to high energies, sufficiently high for nuclear interactions with the stellar atmosphere. On our Sun, for example, electrons, protons, α-particles and heavier ions may be temporarily accelerated to energies up to 10^{10} eV during large solar flares. The temporary high energy particle fluxes in the solar atmosphere are observed directly by radio observations of the Sun (type III, IV and V bursts) and by observation of γ-ray photons during solar flares. The increase of high energy particles following some solar flares (called proton flares) has been recorded by terrestrial instruments. The high energy particles interact strongly with nuclei of the solar atmosphere, as was proven for the first time during the solar flares of August 4 and 7, 1972, when γ-ray lines were recorded by the Orbiting Solar Observatory 7.

Temporary fluxes of high energy particle occur also in the atmospheres of other stars, particularly the magnetic stars. The field intensity and its variations are much larger there than on our Sun. Hence the fluxes of high energy particles on magnetic stars are more abundant and the nuclear reactions more frequent than in the atmosphere of our Sun. The products of such reactions are then easier to detect in the star's spectrum and their amount may be determined. It has been shown, for example, that in the star 3 Centauri A krypton is about 1300 times more abundant than in normal stars and europium in α^2 Canum Venaticorum is even 2000 times more abundant. Other examples of chemical anomalies could be quoted. They may be explained as due to high energy nuclear reactions in stellar atmospheres, where abundant fluxes of high energy particles are produced by variable magnetic fields.

2.1.6. RADIOACTIVE NUCLEI

Besides the thermonuclear reactions in the hot cores of stellar interiors and the high energy nuclear reactions produced by cosmic rays and fluxes of energetic particles, there exists another nuclear process affecting the chemical composition of the universe and, in particular, of our solar system: radioactivity, a spontaneous change of nuclei.

In a chart of nuclides one counts about 1400 nuclides, about 1000 of which have been produced artificially in laboratories. About 270 of the total 1400 nuclides are stable and the rest are radioactive. They change their structure by emission of a positron (β^+ radioactivity), of an electron (β^- radioactivity), of an α-particle (α-radioactivity), or by capturing an orbital electron (electron capture), or by spontaneous division into two parts (fission).

Quantitatively the radioactive decay follows a simple law

$$\frac{dN}{dt} = -\lambda N \qquad (2.3a)$$

or

$$N(t) = N_0 e^{-\lambda t}, \qquad (2.3b)$$

where λ is the decay constant characteristic for each element, N_0 is the number of radioactive nuclides at the time $t = 0$, and N is the number of nuclides at the time t. As may be seen from Equation (2.3b) the period $T = \ln 2/\lambda$ corresponds to $N(T)/N_0 = \frac{1}{2}$. The period T (called the radioactive half-life) is the time required for half of the radioactive nuclei to decay. With time the number of parent nuclides decreases as Equation (2.3b) and the end products increase. The isotope ^{235}U decays into ^{207}Pb and seven α-particles are released in a half-life of 0.7×10^9 yr. Other examples of slowly decaying nuclei are: $^{40}K \rightarrow {}^{40}Ca$ (1.25×10^9 yr); $^{87}Rb \rightarrow {}^{87}Sr$ (4.6×10^{10} yr); $^{238}U \rightarrow {}^{206}Pb$ (4.5×10^9 yr). Comparing the amount of the parent nuclides with the amount of end products we may use Equation (2.3b) as a cosmic chronometer for the age determination of the Earth, Moon and meteorites. The age thus determined is the 'solidification age'.

The relative abundances of radioactive nuclides may be used also for more ancient times viz. for the time scale of nucleosynthesis. The differential Equation (2.3a) becomes more complicated, because nucleosynthetic events in those presolar times increased the abundance of radioactive parent nuclides in the interstellar gas.

$$\frac{dN_i}{dt} = -\lambda_i N_i + P_i f(t), \qquad (2.3c)$$

where i corresponds to a certain species of nuclides, and the term $P_i f(t)$ is the time-dependent production of the nuclides. The relative production of two different species i and j is $P_i f(t)/P_j f(t) = P_i/P_j$, where $f(t)$ is the rate of nucleosynthesis as a function of time. Of course, $f(t) = 0$ after the last nucleosynthetic event (supernova) had influenced the chemical composition of the presolar nebula. As will be shown in Section 5.4, different processes (r-, s-, p-) participated in the nucleosynthesis. Different pairs such as ^{232}Th–^{238}U, ^{187}Re–^{187}Os, ^{235}U–^{238}U and ^{129}I–^{127}I may be used to trace the history of the r-process. Other nuclides are used for the s- and p-processes. The discipline which dates events in the history of the universe on the basis of abundance of radioactive nuclides is called nucleo-chronology.

Summary

Two hadrons (baryons, mesons) interact by nuclear forces (strong interactions). Of the four types of forces between elementary particles, nuclear forces are by far the strongest, but they act only within short distances (10^{-13} cm).

Strong interactions act in the decay of resonance particles, i.e. extremely short-lived baryons and mesons, which decay strongly after about 10^{-23} s into hadrons.

Strong interactions bind nucleons in an atomic nucleus into a unit and ensure its stability. Such interactions have built up most of the existing atoms by thermonuclear reactions in stellar interiors, and during supernova explosions. The liberated binding energy from thermonuclear synthesis is the principal energy source of stars. Practically all the energy on our Earth is of solar origin and it has been released by strong interactions in the solar interior.

Atomic nuclei may be destroyed by strong interaction with high energy particles. Nuclei of atoms in interstellar and interplanetary space, in our atmosphere, on the surface of meteorites, the Moon, etc. are exposed to cosmic rays and are splintered into lighter nuclides. This type of nuclear process, called spallation, leads to fragmentation of heavier nuclei and works against nucleosynthesis in the universe.

Though strong interactions of elementary particles are limited to extremely short distances they are responsible for the chemical aging of the universe and for the energy output of stars.

2.2. Electromagnetic Forces

2.2.1. ELECTRIC CHARGE AND ELECTROMAGNETIC INTERACTIONS

Some elementary particles are endowed with electric charge. Its magnitude is identical for all the particles and it does not change with time. The elementary electrical charge is a universal constant which appears always when electromagnetic interactions are involved. Instead of the elementary charge e a dimensionless quantity $e^2/hc = 1/137$ (called the coupling strength for electromagnetic interactions) is often considered. Charged particles are both the sources of the electromagnetic field and recipients of its effect. While their electromagnetic properties depend on their charge, their mechanical properties depend on their mass. Whereas the electric charge takes only two values $\pm 1.6 \times 10^{-19}$ coulomb, their rest masses have an extended spectrum from 0–1672 MeV (see Figure 1.3). The mass m of a particle is both the source and recipient of gravitation. The gravitation and electrical force decrease with distance r from the particle as m/r^2 and e/r^2. But there are other striking differences between the two forces having their source in the same particle. The electrical force binding an electron to a proton in a hydrogen atom is about 10^{36} times stronger than their gravitational attraction. Under electrical forces two particles attract each other (if unequally charged), repel, or do not affect each other electrically at all. On the other hand, all particles – even with zero rest mass – are attracted to all others by gravitational force. Repulsive gravitational force does not exist.

Electric charge in motion produces a magnetic field. The magnetic force between two moving charges is not really different from the electric force between them. In theory, both electric and magnetic forces are called together electromagnetic forces.

As a result of electromagnetic interactions there is a strong tendency to neutrality in nature. Due to this tendency, the electromagnetic interactions are of secondary importance for the large scale structure of the universe. Gravitation is the main force shaping large cosmic bodies and holding together their systems.

The atom is the smallest structural unit which is neutral. The electrical force of the atomic nucleus holds a family of electrons in their orbits. The complicated interplay of electromagnetic forces, between the nucleus and the electrons and among the electrons themselves, determines the spectrum and chemical properties of the atom. But the positive charge of nuclear protons is not completely balanced by the electron envelope at every place and at every time. The residual represents *chemical forces.* The chemical forces, electromagnetic in nature, are saturated as the atoms unite to become molecules. But even in molecules some small residual electromagnetic forces remain to organize matter into still higher structures (Van der Waals forces).

The weak intermolecular Van der Waals forces are primarily an electrostatic attraction due to the dipole moment, which may be either permanent or induced. The structure continues to crystals (minerals, interstellar grains), from crystals to rocks (meteorites), from rocks to asteroids, satellites and planets; as the mass of the body increases, gravitation acquires increasing importance until it becomes the dominant force (for large satellites and asteroids – see Section 2.4.2).

Although stars are held together by gravitational attraction between the constituent elementary particles (self-gravitation), the electromagnetic interactions of the charged constituent particles are also important for structure and evolution of stars. The rate of energy generation in stellar interiors is controlled by repulsive electric forces between colliding nuclei. The energy generated by thermonuclear reactions in the central parts of a star is transported by radiation towards its surface. This transport consists of a very long chain of alternating emission and absorption processes, i.e., electromagnetic interactions between photons and the stellar plasma. The chain starts with one photon of energy 10^3 to 10^5 eV and ends at the star's surface in the form of thousands of photons with energies 1 to 10 eV. In the gradual degradation of photons the energy is conserved but entropy increases: one calorie in the stellar atmosphere at $T = 10^4$ K has one thousand times more entropy than in the stellar interior at 10^7 K (because $dS = dQ/T$). From the stellar atmosphere large quantities of energy are emitted into the surrounding interstellar space in the form of photons – quanta of the electromagnetic field.

2.2.2. Photon emission

Photons have no charge, although electric charges participate in their creation. The photon has a zero rest mass and moves with velocity $c = 3 \times 10^{10}$ cm s^{-1} in vacuum. Its energy is $h\nu = h(c/\lambda)$. The photon is very readily created by electromagnetic interactions. It is a stable particle because it does not decay. Once born it can move through empty space forever. There are many microwave fossil photons in the universe with age about 10^{10} yr. However photons interacting with matter are easily absorbed and changed into other forms of energy such as heat, chemical energy, kinetic energy (in the Compton effect), photons with the same energy (coherent scattering), photons with smaller energy (fluorescence), electrostatic energy (as in the photoeffect), and rest mass by pair formation, if its energy is sufficiently large.

The lifetime of a photon begins by emission from an electromagnetic interaction. There is a variety of processes by which photons are emitted.

The emission process may occur in a very small volume, as in the decay of a single unstable particle:

$$\pi^0 \to \gamma + \gamma \tag{2.4}$$

$$\Sigma^0 \to \Lambda^0 + \gamma. \tag{2.5}$$

Both electromagnetic interactions occur in a short time ($\sim 10^{-16}$ s in the first case and $\sim 10^{-14}$ s in the second one). Both the processes are a source of γ-rays in interstellar and intergalactic space where pions and hyperons are produced by collisions of cosmic ray particles with atomic nuclei. The energy of photons from Equation (2.4) is about 70 MeV and the photon from Σ^0 decay has about 77 MeV.

Annihilation may be another source of photons, especially of γ-photons. For example π^+ generated by high energy collisions decay as

$$\pi^+ \to \mu^+ + \nu_\mu \quad \text{and} \quad \mu^+ \to e^+ + \nu_e + \tilde{\nu}_\mu \tag{2.6}$$

and the positron annihilates quickly with an electron (Equation (1.15)), producing two 0.5 MeV photons.

An excited atomic nucleus may emit energetic photons. Such γ-rays from radioactive elements are used for therapeutic, industrial, or other purposes by man on Earth. Nucleosynthesis of heavy elements in a supernova explosion (see Section 5.4) results in many neutron-rich nuclei which subsequently decay and radiate γ-photons (mostly 10^5 to 10^6 eV), among other particles. The nuclei excited by high energy collisions de-excite by emitting monoenergetic γ-photons with energy depending on the energy levels in the nucleus. Thus for example, solar γ-photons with 2.23 MeV resulting from deuteron ^{2}H formation have been observed during solar flares.

A common mechanism of photon emission is bremsstrahlung (a German term meaning 'braking radiation'). Electromagnetic radiation is emitted whenever a charged particle is accelerated or decelerated. A well known example is the X-ray radiation emitted by an X-ray tube. Electrons are the lightest charged particles and therefore the easiest to accelerate. Electrons are therefore responsible for the major part of the emission and absorption processes in the atomic world and in the universe. In a plasma, electrons are accelerated whenever they pass near an ion. But a free electron may be accelerated in the field of a neutral atom and, as a result, bremsstrahlung also emitted. This process is important for stellar photospheres with a weak degree of ionization.

The energy of the bremsstrahlung photons comes from the kinetic energy of the electrons. This kinetic energy may be either thermal or non-thermal. The radio emission of a quiet solar corona is an example of thermal bremsstrahlung. On the other hand the bremsstrahlung X-rays observed during an aurora are produced above 90 km in the Earth's atmosphere by non-thermal energetic electrons. In interstellar space γ-photons are emitted by collisions of cosmic-ray electrons with atomic nuclei. In the Earth's atmosphere, the bremsstrahlung photons are also created by a similar process and form the electron–photon component of the secondary cosmic radiation.

Magnetic bremsstrahlung is emitted by electrons gyrating in magnetic fields. In a magnetic field B electrons are accelerated by the Lorentz force and gyrate with frequency

$$\nu = \frac{eB}{2\pi m_e c}. \tag{2.7}$$

The gyrating electron emits radiation of the same frequency. If, however, the gyrating electrons are relativistic, the emission is not monochromatic, as would correspond to formula (2.7). Many harmonic components of the gyrofrequency occur and the emitted radiation is continuous. The radiation – called synchrotron radiation – is polarized and is emitted only in a narrow cone in the direction of the electrons' flight. The spectrum of the synchrotron source may range from radio waves to X-rays, as in the case of the Crab nebula. Jupiter's decimetric emission, some radio bursts on the Sun (e.g. type IV), and radio emission of galaxies are other examples of synchrotron emission of relativistic electrons in extended cosmic magnetic fields. The shape of the synchrotron spectrum differs markedly from thermal radiation and the former is polarized. The non-thermal radio sources may be thus easily recognized.

The isotropic microwave radiation (fossil, relic or 3 K radiation) is generally considered to be very old, as old as the universe itself. It is a thermal radiation which represents vestiges of the Big Bang (Section 5.2). Its microwave photons fill up the entire universe. In collisions with relativistic electrons of cosmic radiation the energy of the photons may be substantially increased. This energy transfer from a relativistic electron to a microwave photon (inverse Compton effect) is a source of γ-radiation, especially in the vast intergalactic space.

Atoms, ions, molecules or crystals (i.e. systems of elementary particles) emit radiation after having been excited. By deexcitation or recombination, photons of very different energy may be created, ranging from Ångström to decimeter wavelengths. The transition of an electron to the innermost shell in heavy atoms or ions produces very energetic or 'hard' photons. Such is the case in hot condensations in the solar corona. On the other hand, transitions between high energy levels of atoms and molecules result in cm or dm radiation. In general a transition between two energy levels of a certain species of atom or molecule results in a characteristic energy of the emitted photons. By measuring the spectral lines, astrophysicists get information among other things about the chemical composition and state of excitation and ionisation of the plasma in the universe.

The recombination of ions produces a spectral continuum with a sharp edge and decreasing intensity to higher frequencies. Recombinations of ionized atoms represent an important emission process in stars: H and He recombine in atmospheres of hot stars, metals in cool stars. Free electrons recombine also with neutral atoms and emit photons, e.g.

$$H + e^- \rightarrow H^- + h\nu. \tag{2.8}$$

The *formation of negative hydrogen ions* is an important emission process in the solar photosphere and in the atmospheres of many other stars. The photon $h\nu$ carries away the ionization energy (0.75 eV corresponding to 16 550 Å) plus the kinetic energy of the captured electron. The free electrons have different energies according to the Maxwell–Boltzmann distribution law so that the photons emitted by Equation (2.8) form a continuous spectrum, with the long-wave end at 16 550 Å. Most of the light photons from the Sun are emitted as given by Equation (2.8).

The *reversal of electron spin* in H I atoms is a photon emitting process important in interstellar space only, where collisions are very rare. A H I atom with the electron spin parallel to the proton spin has a slightly higher energy than in the antiparallel orientation. The atom gets rid of this tiny energy surplus after about 10^6 yr by reversing the electron spin and emitting a photon of frequency 1420 MHz, corresponding to a wavelength of 21 cm. The radiation of this wavelength brings important information about the distribution of neutral hydrogen in our Galaxy and other galaxies as well.

Čerenkov radiation should be mentioned more for the sake of completeness than for its importance in the universe. When a high energy charged particle passes through a transparent medium and the velocity v is greater than that of light in the medium ($v > c/n$, if n is the refraction index), bluish light is emitted. It is an electromagnetic shock wave induced in the medium by the passing particle and is an optical analog of the sonic boom. Flashes of Čerenkov radiation are produced in our atmosphere by charged cosmic ray particles.

2.2.3. PHOTON UNIVERSE

Considering the role of all the elementary particles (Figure 1.3) one comes to the conclusion that nucleons, pions, electrons and photons are the most important particles for the structure and evolution of the universe.

Photons transport energy and carry information from one place in the universe to another. Practically all our knowledge about the universe has been gained by analysis of photon radiation. The contribution of non-photonic information (such as corpuscular radiation, meteorites, lunar samples, direct space research *in situ*) is relatively small.

To extract information from the photon radiation astronomers measure its different properties – flux or intensity (photometry), direction of incidence or position of the source (astrometry), polarization (polarimetry), spectral composition (spectrophotometry) and time variations of the properties. From the measured data distances of celestial bodies, their size, mass, chemical composition, temperature, pressure, magnetic fields, rotation, motion, and other characteristics may be deduced. Additional information brought by photons concerns the medium between the observed body and ourselves (e.g. interstellar matter, interstellar magnetic fields, interplanetary matter, terrestrial atmosphere etc. (Figure 2.3)).

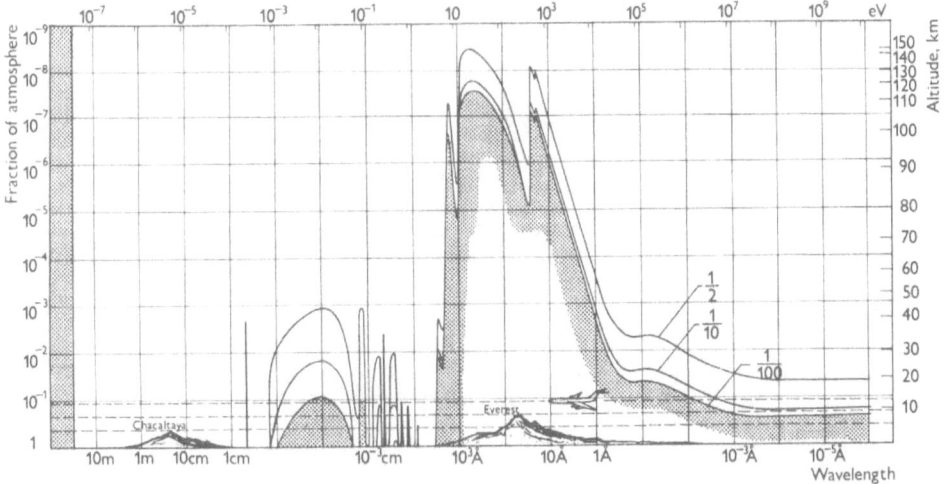

Fig. 2.3. Attenuation of electromagnetic radiation by the terrestrial atmosphere. Abscissae are wavelengths, ordinates represent fraction of atmosphere or altitude in kilometers. The numbers 1/2, 1/10, 1/100 indicate how much from the original radiation flux (at the top of the atmosphere) passed through. We see for example, that for the extreme UV (a few hundreds of Å) the terrestrial atmosphere is opaque, while for meter wavelengths it is transparent.

Our Earth is irradiated by photons arriving from all directions. The amount of their energy in a certain frequency range incident from a unit solid angle on (or crossing) unit area per second is called intensity of radiation $I(\text{erg cm}^{-2}\,\text{s}^{-1}\,\text{sr}^{-1}\,\text{Hz}^{-1})$. It depends upon the direction of arrival (α, δ), upon frequency (ν), upon polarization P, and also upon time (t):

$$I = I(\alpha\delta, \nu, P, t). \tag{2.9}$$

A precise determination of the direction of incident photons i.e. of position on celestial sphere (α-right ascension, δ-declination) has always been one of the basic tasks of astronomy. The local inertial frame of reference is fixed by absolute positions and proper motions of fundamental stars (FK4 – Fourth Fundamental Catalogue). The knowledge of accurate values of precession, nutation and aberration is necessary. The positions of compact radio sources may be determined by means of radio interferometers with higher accuracy than in absolute optical measurements.

Enhanced radiation (higher intensity) from certain directions (discrete sources) is superposed on background or diffuse radiation coming from extended areas of the sky. It has been detected over 16 decades of ν from radio waves of about one MHz up to 3×10^{16} MHz radiation i.e. γ-photons with energies 100 MeV. One component of the diffuse radiation comes from our Galaxy – mainly from its disk – and is therefore concentrated towards the galactic plane. The other component of the diffuse radiation is the isotropic background radiation, which is truely cosmic because it arrives from very distant extragalactic sources and from intergalactic space (Figure 2.4). The isotropic background radiation permeates the vast spaces between galaxies and provides information on the history and structure of the universe.

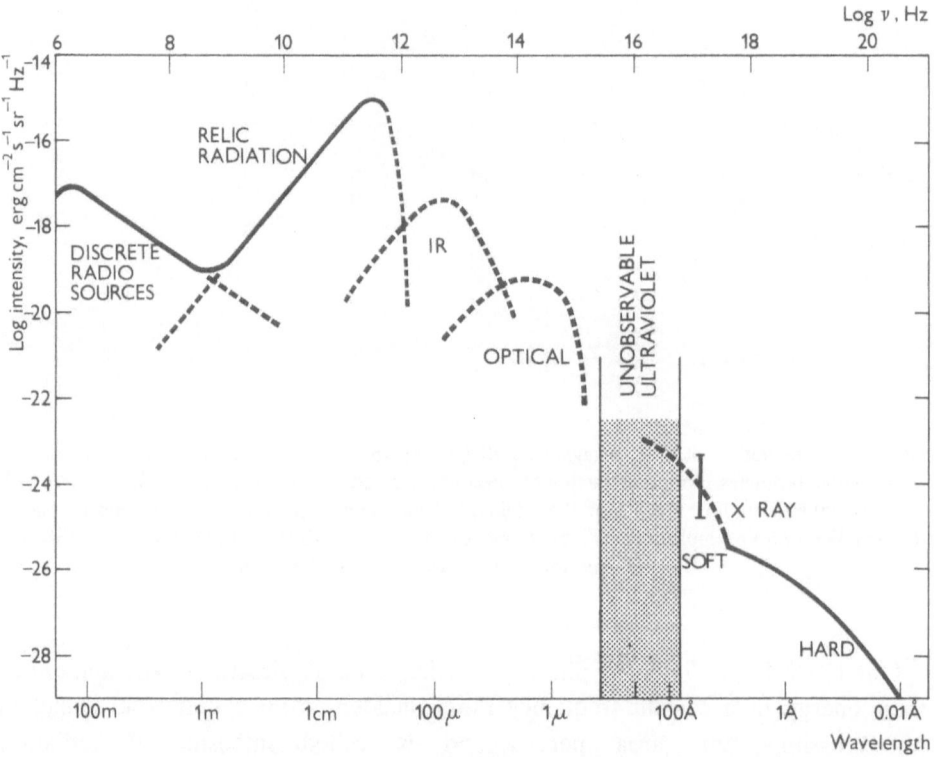

Fig. 2.4. Spectrum of the isotropic background radiation of the universe (Longair, Sunyaev). Full line shows observational results, dashed line theoretical estimates.

The second independent variable in Equation (2.9) is frequency ν of the radiation or energy $h\nu$ of the incoming photons. The photons from the universe may have very different values, ranging from about 10^{-11} eV (e.g. decametric radiation in the type III bursts from the Sun) up to 10^{17} eV (e.g. cosmic-ray γ-photons). It is understandable that measurements in this huge energy range, covering about 28 orders of magnitude, must use a variety of techniques. Radio, submillimetre, IR, UV, X-ray, and γ-ray ranges of the cosmic photons have been disclosed to man, thanks to modern technology, and an enormous new world has been revealed to him. The extension of the visible spectrum on both sides – to higher frequencies (UV, X, γ) and to low frequencies (IR, submillimeter, radio) – meant a revolution in our understanding of the structure and evolution of the universe. Radio galaxies, microwave background, quasars, pulsars, X-ray sources, IR sources, γ-ray bursts, solar radio burst, and Jupiter's radio emissions are some of the objects and events about which an optical astronomer even a quarter century ago had not the slightest idea at all.

The dependence of I on ν, i.e. spectrum, contains important information on its source. The spectrum may cover one or more spectral regions (radio, submillimeter, IR, visible, UV, X, γ – see Figure 2.5). The distribution $I(\nu)$ may be continuous, as in synchrotron or black-body radiation; it may be limited to certain ν only, as in line spectrum and molecular spectrum; or it may be a combination of both. A source recorded in one spectral region need not be recorded in the other regions. A radio source may also be a source of X-ray photons or light photons, but not necessarily. The appearance of the sky $I_\nu(\alpha\delta)$ for constant ν, P and t depends on the frequency ν (or the frequency range) chosen for the sky survey. The radio sky is different from the X-ray sky and both differ from what one observes in light, though some sources may be identified on two or all three. This will be discussed in Sections 2.2.4 to 2.2.9, where the appearance of the sky in different spectral regions will be described.

The symbol P in $I(\alpha\delta, \nu, P, t)$ concerns the alignment of electromagnetic vibrations in the plane perpendicular to the direction (α, δ). The electric vector of the radiation from many sources is not randomly distributed but shows a preferential direction (azimuth of polarization). The radiation may be linearly, elliptically, or circularly polarized, and the fraction of the polarized component may vary from zero (unpolarized radiation) to one (complete polarization). The symbol P is used to represent the azimuth, character and degree of polarization of the radiation intensity I. Let us mention a few examples of sources of polarized radiation: light scattered by our atmosphere, OH and H_2O interstellar clouds emitting highly polarized radiation by a maser mechanism, type IV bursts in the solar corona, remnants of supernovae, light of many stars is partially polarized with the electric vector parallel to the local spiral arm in which the stars are embedded, radio galaxies, jets from galactic nuclei, etc.

The time in $I(\alpha\delta, \nu, P, t)$ is the independent variable. The appearance of the sky $I(\alpha, \delta)$ (for ν and P constant) changes with time: either power of the source changes (variable stars, supernovae, etc.) or its position $(\alpha\delta)$ does (as in the motions of planets, proper motion of stars, parallax, deflection by gravitational field, scintillation, refraction etc.). The spectrum $I(\nu)$ of many sources is also time-dependent (the solar spectrum and in particular its radio and X-ray portions are strongly variable with solar activity; softening of the spectrum of some radio sources; spectral changes in some variable stars and supernovae during short time scales, and changes of stellar spectra during their evolution i.e. during long time scales are other examples).

The time in Equation (2.9) is the photon arrival time, the time of observation. This is the same for all the arriving photons, without distinction from where they arrive: a

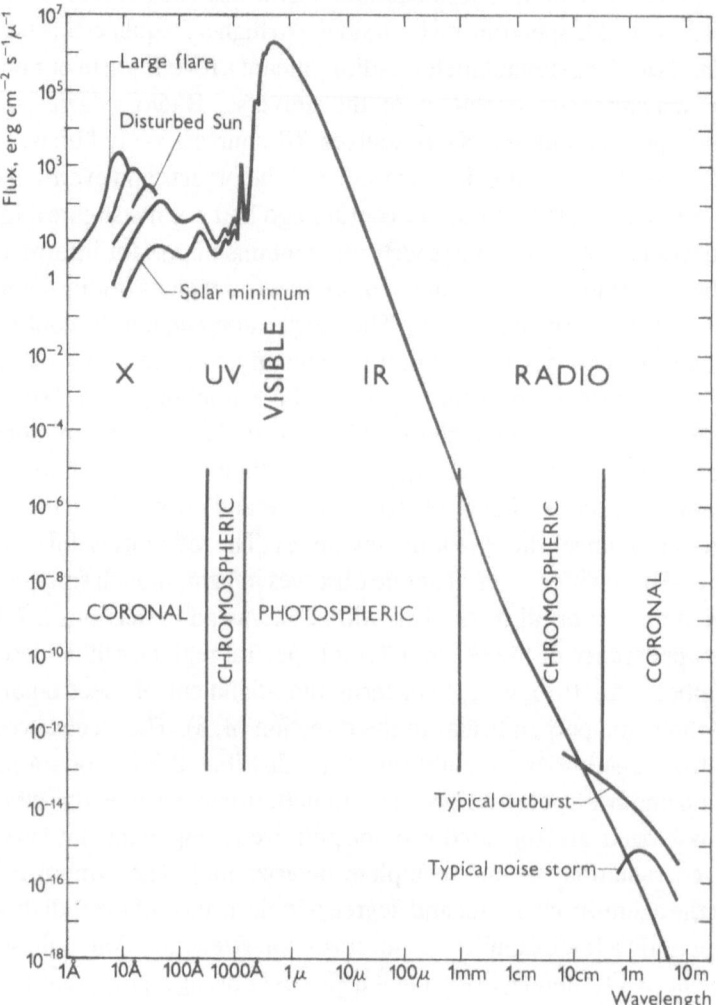

Fig. 2.5.   The best known source of radiation for us is the Sun. It emits electromagnetic waves in a broad range, from gamma rays to long radio waves. The curve represents the incident radiation flux (at the top of terrestrial atmosphere) from the whole Sun for each wavelength. The gamma, X, UV and radio radiations are variable with time (as a result of solar activity), while the visible and infrared fluxes remain constant. The solar spectrum is a synthesis of radiations from the whole solar atmosphere – as indicated in the figure.

microwave photon from the Moon 1.3 s old and a fossil photon 10^{10} yr old are both observed at the same time t. But even radiation from the same source emitted at the same moment (e.g. the flash of a pulsar) arrives at different times when observed at different frequencies, due to the fact that the velocity of propagation depends on wavelength (dispersion).

2.2.4. RADIO SKY

(a) *Background Radiation and Radio Sources*

The background radiation of the sky at $50 \text{ cm} < \lambda < 300 \text{ m}$ has two components. There is an *anisotropic component* which is highly concentrated to the galactic plane. It is emitted by relativistic electrons gyrating in interstellar magnetic fields (synchrotron radiation) of our Galaxy. There is also an *isotropic component* of the background which is also non-thermal but it is of extragalactic origin. A direct proof exists that part of the diffuse background radiation must be of extragalactic origin in that the radio brightness of the sky is diminished in the direction of the nebula designated as 30 Doradus. The nebula is in the Large Magellanic Cloud and the brightness diminution, observed at λ 15 m, is caused by absorption of radio radiation arriving from yet further distances than the Large Magellanic Cloud, viz. of cosmic isotropic background.

The spectrum of the isotropic component, measured at different λ in the direction to the galactic poles, suggests that the isotropic background is integrated radiation of many discrete, very distant, and therefore unresolved extragalactic radio sources (normal galaxies, radiogalaxies, and quasars). Character of the isotropic background spectrum corresponds to the non-thermal spectra of those discrete sources. However, the present density of such sources in the near-by universe seems insufficient to explain the observed intensity of background radiation at $\lambda > 50$ cm. This is probably an effect of evolution: the number of powerful radio sources in the past (i.e. at large distances) was much larger than at present. But the isotropic component is probably integrated radiation of very distant sources.

The intensity of the background spectrum for $\lambda > 50$ cm decreases with decreasing wavelength, because it is non-thermal radiation. However in the region $300 \ \mu m < \lambda < 50$ cm the intensity of the extragalactic background radiation is substantially increased and has a Planckian distribution corresponding to 3 K. It is believed to be relic radiation from the earliest phases of the universe, when more energy was in the photons of the background than in particles with rest mass. The photons coming isotropically from extragalactic space are called relic, fossil, black body, or 3 K background radiation. They are the oldest photons of all. Emitted during the radiation era of the universe, they survived until present times. Only their energy has been substantially reduced by expansion of the universe (Section 5.2).

Superposed on the diffuse background are compact radio sources, each a few minutes or less in extent. Some forty thousand have been catalogued to date. Only a few hundred of the more intense sources have been studied in detail. The optical counterpart of some radio sources have been identified and the distances to the sources determined. There are sources in our Galaxy and many powerful sources in the extragalactic universe, associated with optical galaxies and quasars (Figure 2.6). Bodies in the solar system are also radio sources, weak but observable due to their proximity.

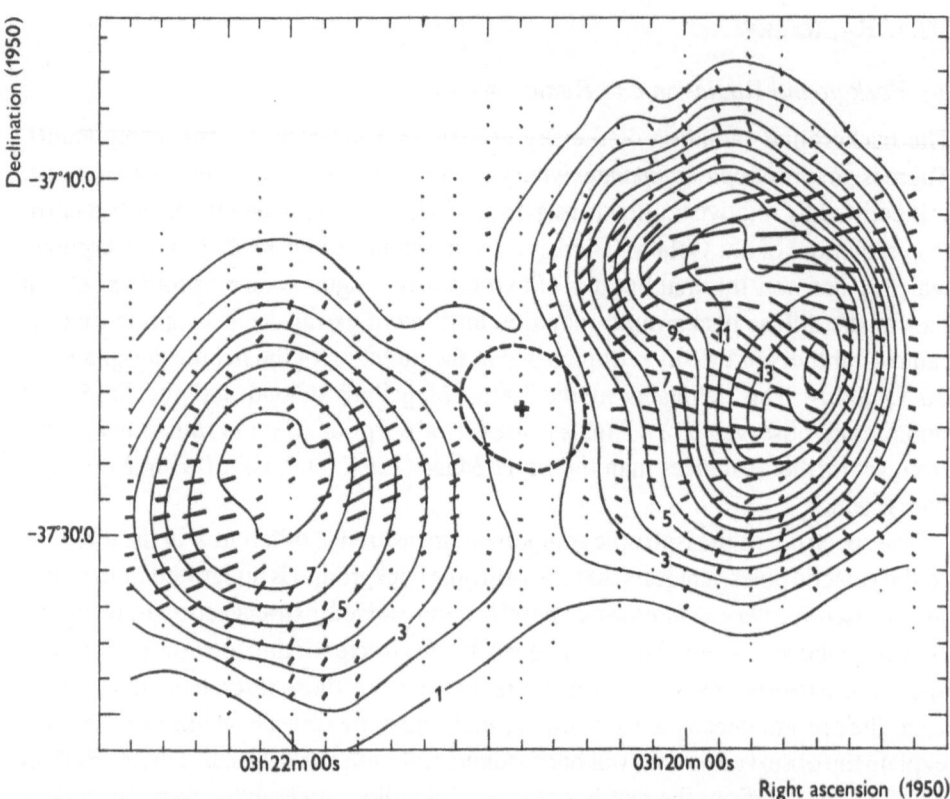

Fig. 2.6. A radio source on wavelength 6 cm. Intensity distribution and linear polarization in the radio galaxy Fornax A. The isophote interval is 0.155 K. The position of the parent optical galaxy (NGC 1316) is marked by a cross and its outer limits by a broken line (CSIRO – Division of Radiophysics).

Catalogues of both galactic and extragalactic sources list the name of the source, its position (α, δ), and intensity or flux, which is the power received from the source per unit area and unit frequency. The magnitude is called flux density and is usually measured in flux units defined as

$$1 \text{ f.u.} = 1 \text{ Jansky} = 1 \text{ Jy} = 10^{-23} \text{ erg cm}^{-2} \text{ s}^{-1} \text{ Hz}^{-1} = 10^{-26} \text{ W m}^{-2} \text{ Hz}^{-1}. \quad (2.10)$$

In solar astronomy a larger unit, solar flux unit (s.f.u.) is used, where 1 s.f.u. $= 10^4$ f.u. Other data listed in catalogues of radio sources are the size of the source and its distance and optical identification.

(b) Radio Sun and Radio Planets

The Sun was the first observed discrete radio source. Compared with other radio sources, its power is very weak. As it is near by, the flux of its radio waves is strong enough to be studied even with modest instruments. The Sun is the best known radio source in the sky. Its spectrum and intensity distribution over the disk is recorded daily. Variations in time of both are followed by dynamic spectrographs (Figure 2.7) and radioheliographs.

As may be seen from Figure 2.5, the quiet Sun is a thermal source. The radiation is mainly emitted from coronal condensations i.e. hot and dense plasmas over sunspots. The latter are regions in the photosphere with strong magnetic fields 500 to 5000 G. The thermal electrons (10^2 eV) lose their energy by bremsstrahlung, either in the field of positive ions or by gyrating in the magnetic fields.

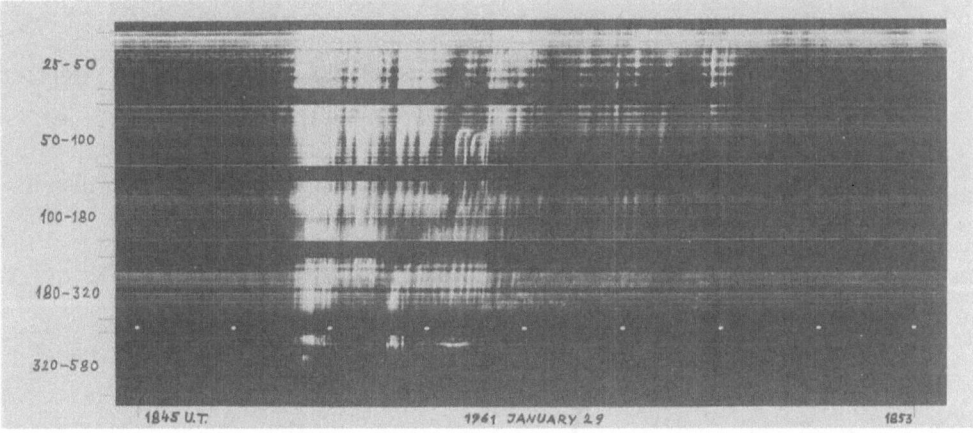

Fig. 2.7.   Radio spectrum of the Sun from 1961, January 29, between 18.45 UT and 18.53 UT. Abscissae represent time, ordinates frequency from 580 MHz (bottom) to 25 MHz (top) (CSIRO – Radiophysics Division). The radiation is called Type III bursts and it is excited by fast particles ejected from the Sun.

During solar activity, especially during flares, the solar atmosphere radiates temporarily high fluxes of non-thermal emission (noise storms, outbursts, bursts). High energy electrons accelerated by flares radiate by synchrotron radiation (type IV burst), excited plasma waves convert part of their energy into radio waves, giving rise to slowly drifting type II and fast drifting type III bursts. Radio bursts are very spectacular, especially on meter wavelengths; there the intensity (or flux) may fluctuate a thousand fold within a few seconds. For radio waves a magnitude called brightness temperature T_b is often used instead of intensity:

$$I_\nu = 2kT_b/\lambda^2 \qquad (2.11)$$

(Rayleigh–Jeans law). The brightness temperature of the radio bursts rises very fast from 10^6 K to 10^{11}–10^{12} K. The fluxes of the radio photons at meter wavelengths may increase a million times or more within seconds, while the optical emission remains practically unchanged.

Planets emit thermal radiation the spectrum of which depends primarily upon the temperature of the planetary surface. Since $I \propto \lambda^{-2}$, the measurements are made at the shortest possible wavelengths. Thus, for example, Venus has been measured at different wavelengths between 0.1 cm and 70 cm. The brightness temperature of its dark hemisphere (measured during inferior conjunctions) is rather high, about 600 K for $3\,\mathrm{cm} < \lambda < 50\,\mathrm{cm}$, although the clouds are known to be at about 230 K. For $\lambda < 3\,\mathrm{cm}$ the intensity decreases due to absorption in the planets atmosphere. The distribution across the disk and polarization near the limb indicate that the radiation is emitted by a solid surface. This surprisingly high temperature measured from the Earth was recently confirmed by measurements from space probes.

The other planets emit radio radiation like normal thermal sources.

(c) *Radio Sky at Different Wavelengths*

In mm and cm wavelengths Venus is the brightest planet. The Moon is 10^2 brighter and the Sun about 10^4 times brighter than Venus. In dm wavelengths the picture is changed: Jupiter becomes the brightest planet, because its non-thermal emission from its radiation belts substantially (about one order of magnitude) prevails over its thermal component (Figure 2.8).

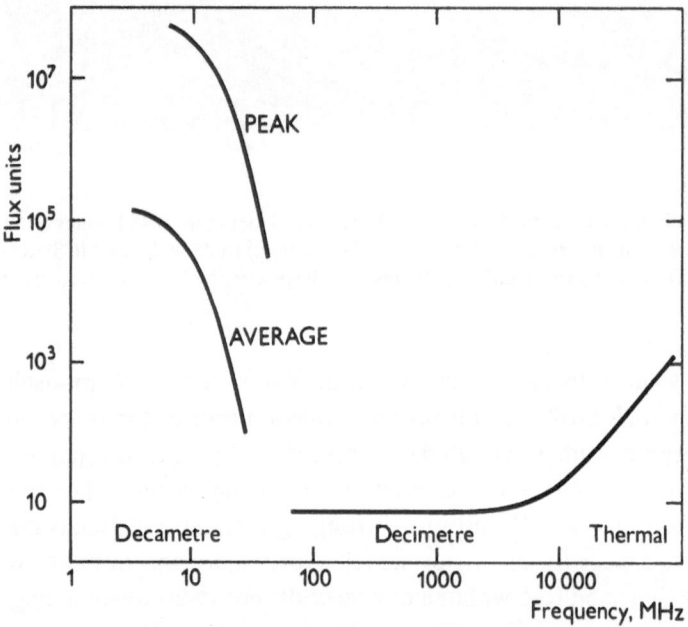

Fig. 2.8.   Schematic representation of Jupiter's radio spectrum (according to Carr, Newburn). The flux scale (ordinates) is in flux units (1 fu $= 10^{-26}\,\mathrm{W\,m^{-2}\,Hz^{-1}}$) corresponds to the planet opposition (about 4 AU). The thermal emission ($> 10\,000$ MHz) originates in Jupiter's atmosphere (effective temperature of 134 K). The decimeter radiation is linearly polarized, its source has a large angular extent – it is synchrotron radiation from the radiation belts in the Jovian magnetosphere. The decametric radiation is strongly variable, it is non thermal but not yet completely explained.

At cm and dm wavelengths some radio sources outside the solar system become brighter than Venus and Jupiter (Crab nebula, Omega nebula, Cas A).

At meter wavelengths our Sun becomes comparable with the brightest sources like Cas A or Cyg A. The reason is simple: with increasing wavelength the intensity (flux) of the thermal solar radiation decreases while the intensity of the non thermal sources increases. Hence in the decametric sky the quiet Sun is quite faint and some non-thermal sources (Cas A, Cyg A, galactic center) are much brighter. Sporadically even Jupiter may become brighter than our Sun, when strong non-thermal bursts are excited in the giant planet's magnetosphere.

(d) Galactic Radio Sources

The earliest optical identifications of radio sources in our Galaxy were made with supernova remnants and ionized hydrogen clouds around bright O and B stars (H II regions). The Crab nebula is the best known radio source. It contributed much to our understanding of the universe. Its relativistic electrons emit polarized synchrotron radiation with a spectrum ranging from radio to X-ray radiation.

In recent times radio emission from some *active stars* has been detected and observed. The thermal radio emission of normal stars is below the sensitivity limit of present radio telescopes. But the situation is different for stars with some type of activity by which non-thermal radiation may be emitted. In a manner similar to solar flares, the radiation of *flare stars* may increase considerably by a factor 10^6 or more. Radio bursts from flare stars have actually been detected by patient collaboration of optical and radio astronomers.

Large radio telescopes have recently detected radio emission from *red giant* stars (e.g. Betelgeuze which is probably a seat of extensive prominence activity). *Blue companions* in some binary stars (Antares, Algol, β Lyrae, and the star associated with Cyg X-1) are also radio emitters. There is spectroscopic evidence for streamers of plasma between components of binaries. Plasma streamers from one star to the other are capable of generating radio waves and even X-rays. There are, however, close binary systems which are radio quiet.

Novae have been shown to be strong emitters of radio waves. After the outburst of Nova Delphini 1967 and Nova Serpentis 1970 their radio emission was observed to be steadily increasing, probably from the ejected shells of plasma.

The X-ray source Sco X-1 is a triple radio source – one source at the position of Sco X-1, which is also visible optically, and the other two on either side of it. It is a very variable radio source, changing its intensity by a factor of 60 in less than one day. Another well known X-ray source discovered by the UHURU satellite is Cygnus X-3. Although not visible optically, it is a radio source with a violent flare activity. Its distance, determined from absorption of radio waves in the interstellar medium, is about 11 000 pc which is greater than the distance to the galactic center.

The term *radio star* may be used again after having been discarded for several years.

Pulsars are a special class of variable radio stars. They are very small neutron stars (see Section 5.4) with very fast rotation. The pulses are repeated with great regularity. While the repetition of the pulses is very regular, their intensity varies in an irregular way, showing variations ranging from seconds to years. The period of the pulses P is different from one pulsar to the next. For the known pulsars (about one hundred) P lies within the interval $0.033\,\mathrm{s} \leqslant P \leqslant 4\,\mathrm{s}$, and the duration of the individual pulses is in the interval from 0.005 to 0.050 s. It seems that the young pulsars rotate rapidly while the old ones rotate more slowly. It has been observed that most of the pulsars are slowing down very uniformly. The period increase $\dot{P}$ is different for different pulsars (for the Crab pulsar it is 4×10^{-13} s s^{-1}, while for some other pulsars it may be as small as 10^{-16} s s^{-1}). Several pulsars (e.g. the Crab pulsar) have in addition shown sudden decreases in the period of the pulses, which then resume their normal slowing down. It is supposed that such sudden changes are caused by star quakes of the pulsar. The regular repetition of the pulses is due to the fast spinning of the neutron star. The mechanism producing the emission is not yet quite understood, although it is probably connected with the very strong magnetic fields of the pulsars. The pulses can be received at all wavelengths, from a few meters to a few centimeters. The Crab pulsar, which probably feeds relativistic electrons into the Crab nebula, has the shortest known period (0.033 09 s). It is a variable source, not only in radio radiation but also in light, X-rays, and γ-rays.

(e) Extragalactic Radio Sources

The radio sources outside our Galaxy have been optically identified with normal spiral galaxies, Seyfert galaxies, elliptical galaxies which are often the brightest member of a cluster, elliptical galaxies with an extended halo, dumbbell galaxies with a double nucleus imbedded in a common halo, N-galaxies with a compact bright nucleus superimposed on a faint nebulous envelope, and quasars. Roughly speaking, the radio luminosity of the source increases in the order of the above sequence.

The synchrotron mechanism is mainly responsible for the radio emission of the extragalactic sources. At wavelengths longer than 10 cm almost all the continuous emission from galaxies is synchrotron radiation. Their thermal radiation is too weak to be observed, except from their giant H II regions (e.g. in M 33).

Many extragalactic radio sources show a double structure: two extended radio-emitting regions situated symmetrically about the nucleus of a galaxy or a quasar. The total extent of the double source is somewhere between 35 kpc (as for 3 C 31) and 750 kpc (Centaurus A). Detailed maps of the sources show a high complexity, with small-scale structures, emission bridges, more double sources along the same axis, etc. About one fifth of the well resolved sources show a spherically symmetric halo distribution about a core. In more complicated cases a core-halo source may contain also a double source in the form of an optical jet and counterjet, as in M 87.

The amount of energy stored in the radio galaxies and in quasars (in the form of relativistic electrons, ions and magnetic fields) is very high (see Section 4.3). It is now generally believed that the energy is released by explosions of the galactic nuclei. The two symmetrical radio sources or optical jets are apparently remnants of an immense explosion from the nucleus of the optical galaxy. The masses and fields thrown out contain many energetic particles (cosmic rays) which radiate the synchrotron fluxes. In the last years more evidence has accumulated that eruptive activity of galactic nuclei accompanied by strong synchrotron emission is a common phenomenon – even in galaxies that are otherwise normal.

(f) *Radio Line Emission*

The radiation of the sources discussed above – whether thermal or sychrotron – has a continuous spectrum. In recent years sources of line spectra have also been discovered by radio astronomers. The lines are either in emission or in absorption when observed against a distant source with continuum synchrotron spectrum (e.g. in the direction of a supernova remnant such as Cas A).

The H I in interstellar space emits or absorbs radiation of frequency 1420 MHz (i.e. $\lambda = 21$ cm). This line is due to the hyperfine-structure transition of the ground state of hydrogen atoms (i.e. flipping over of the electron spin). The observed profile of the 21 cm line depends on the direction of observation. It usually has several maxima which are shifted in wavelength. The maxima correspond to different clouds of H I moving with different radial velocities. The distribution of the 21 cm line in the sky shows that H I is concentrated to the galactic plane and forms a flat disk. The rotation of the Galaxy results in clouds having different radial velocities when observed from the Earth, depending on their direction and distance from us. The rotation of H I clouds around the galactic center depends on their distance from the center (differential rotation). The dependence of rotation velocity upon the distance from the galactic center is known. It means that the radial velocities of H I clouds, measured from the 21 cm profiles, can be converted into distances of the clouds (Figure 2.9). This method permits the determination of the distribution of neutral gas in our Galaxy to substantially larger distances than can be done by optical astronomers. Since H I is concentrated mainly in spiral arms, the two-dimensional 21 cm maps of the sky have lead to the three-dimensional spiral structure of our Galaxy. Thus radioastronomy presents a picture of our own Milky Way from outside, as it looks from extragalactic space.

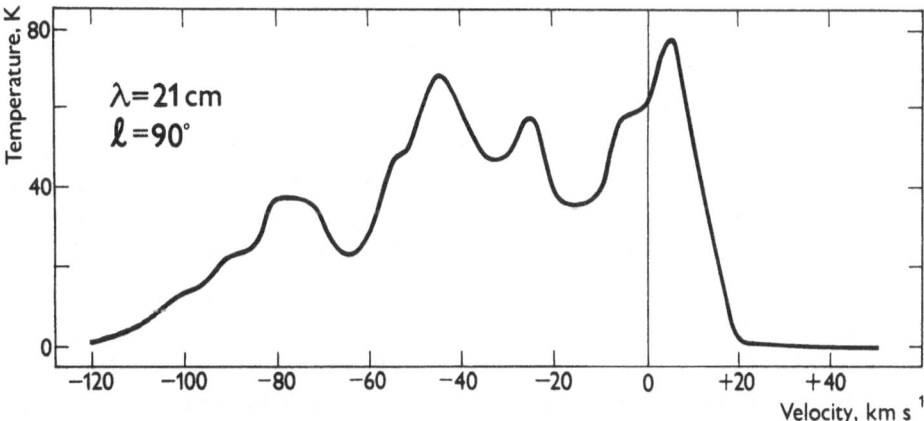

Fig. 2.9.   Example of the 21 cm interstellar emission. The abscissae are radial velocities, the ordinates are intensities in temperature scale. The individual peaks are emitted by neutral hydrogen in distant spiral arms of our Galaxy. The rotation velocities of different parts of the Galaxy are known and the measured radial velocities can be transformed into distances of the arms from us.

Ionized hydrogen regions emit radio recombination lines which result from recombination and subsequent transitions between levels with very high quantum number (about 100 or more). The lines emitted by electron transitions to the n-th level are called n_α, n_β, n_γ ... lines. In H II regions a large number of n_α lines have been observed for many numbers n and over a large frequency range. Some lines n_β, n_γ, n_δ, n_ε have also been observed.

Analogous lines of He II and other elements have been found in the radio spectra of bright diffuse nebulae. These nebulae are clouds of interstellar plasma near or around hot stars. The UV radiation of the star excites and ionizes atoms of the nebula. By de-excitation and recombination different lines and continua in the UV, visible, IR and radio spectral regions are emitted. Due to the relatively high ionization of the most abundant element – H – the line emission nebulae are called H II regions (or Hα-emission areas). Their radio radiation is of thermal origin.

Many H II regions were also identified as He II regions. For example, the Orion nebula M 42, Lagoon nebula M 8, Omega nebula M 17 and many others. The relative intensities of the H and He recombination lines are a good indicator of He abundance in the galaxy relative to the H. It is a very important ratio, because it may decide whether the He was formed only by nucleosynthesis in stars or whether its major part had already been synthesized during the Big Bang. The present value of N_{He}/N_H is about 0.07 to 0.1. There are however observational difficulties (blending of He lines by C lines; the ratio depends on the temperature of the exciting star) so that more precise observations are still needed to decide what is the value of the He abundance.

During the last decade many molecules have been discovered to exist in interstellar gas, from diatomic hydroxyl OH to sevenatomic acetaldehyde CH_3CHO. Some of them emit millimetric waves, such as CO (2.6 mm), CN (2.6 mm), methylacetylen CH_3CCH (3.5 mm); others radiate in cm wavelengths, such as H_2O (1.3 cm), NH_3 (1.3 cm), methanimine CH_2NH (5.7 cm); while others radiate in dm-wavelengths, such as acetaldehyde CH_3CHO (20.0 cm), formic acid CHOOH (18.0 cm), methyl alcohol CH_3OH (36 mm), etc.

The radical OH has been observed in different directions in our Galaxy, either in absorption or in emission. The ratio of the emitted OH lines, their sharpness, polarization and the extremely small size of the emitter ($<10^{14}$ cm, smaller than the solar system) indicate that the OH sources must be a cosmic maser. The strongest OH line is twice as strong as the strongest H line, although H is 10^7 times more abundant than the OH molecules. The small OH clouds act as a giant maser, picking up a very weak signal and amplifying it immensely. The OH masers are associated with Mira stars and with dust clouds near H II regions; they are variable with periods of weeks or months, and their luminosities may be as high as $L_\odot$. The OH and H_2O regions have sizes and densities intermediate between typical interstellar clouds and stars or planets. They are plausibly associated with protostars or primitive nebulae. OH emission has been detected also from other galaxies, e.g. M 82.

As one can see, the universe is a natural transmitter of radio waves, which transmits power far beyond the capabilities of man, in very broad frequency ranges through times measured in thousands of millions of years. Unfortunately, radio communication, television, radio, aviation and many other human activities pollute the natural radio emissions on our planet to such a degree that special international conferences had to recognize the radio astronomy as a communication service and emphasize the need for complete radio quiet at least at some frequencies. However, the radio spectrum is heavily used and is being rapidly filled by new users. Moreover man, with his daily activity heavily pollutes the natural radio environment. For example, the radio noise measured in 1972 above certain small American cities corresponded to brightness temperatures up to 10 000 K (in the range 73 MHz to 440 MHz).

2.2.5. INFRARED SKY

Infrared radiation from the universe is strongly absorbed by molecules of H_2O and CO_2 in the lower part of the Earth's atmosphere, so that the IR observations from the ground are limited to unabsorbed wavelengths (windows). For complete observations of the IR sky in wavelengths $1 \ \mu m < \lambda < 300 \ \mu m$ (10^{12} to 10^{14} Hz), the instruments have to be placed at high altitudes, e.g. above the atmospheric absorption. See Figure 1.3.

(a) *Planets and Sun in Infrared*

Planets are strong IR sources and their spectra with molecular bands of NH_3, CH_4, H_2O, and CO_2 have been studied by Fourier spectroscopy, which is a powerful tool to obtain high-resolution spectra of the IR radiation. Such spectra are a rich source of information about chemical composition, pressure, and temperature distribution in the Earth's atmosphere. The IR radiation of the Earth is measured from satellites to determine the geographical and altitude distribution of some physical parameters of our atmosphere.

The transition zone between the Sun's photosphere and chromosphere emits radiation of wavelengths about $100 \ \mu m$ (0.1 mm). Shorter wavelengths $\lambda \leqslant 10 \ \mu m$ originate in the photosphere, while longer wavelengths come from the chromosphere. The deepest layers of the solar atmosphere are observed in wavelengths about $1.6 \ \mu m$.

(b) *The Central Region of the Galaxy*

Infrared photons penetrate interstellar space much more easily than does light. The dust between the galactic center and the solar system absorbs light nearly completely, to a factor 10^{-10}. On the other hand, the IR radiation is attenuated substantially less so that the structure of the central regions of our Galaxy may be studied and compared with other galaxies.

The IR central region of our Galaxy coincides with the central radio source Sagittarius A. The size and power depends upon the wavelength of observation (1° for 1 to 5 μm, 15 arc s for 3 to 20 μm; 2°×4° in 100 μm radiation). The IR luminosity of the central region in 80 to 120 μm is $\sim 10^9$ $L_\odot$ which is about 1% of the total luminosity of the Galaxy. The IR emission is not smoothly distributed but has a patchy structure.

(c) *Infrared Stars*

Many optical stars and galaxies are rather weak in IR radiation. But there exist stars, nebulae and galaxies which are powerful IR sources. Some cosmic objects are observable only in the IR. For example, Becklin's object in the Orion nebula (Figure 2.10) emits mostly at λ 5 μm, remains undetected in light, and resembles a 600 K

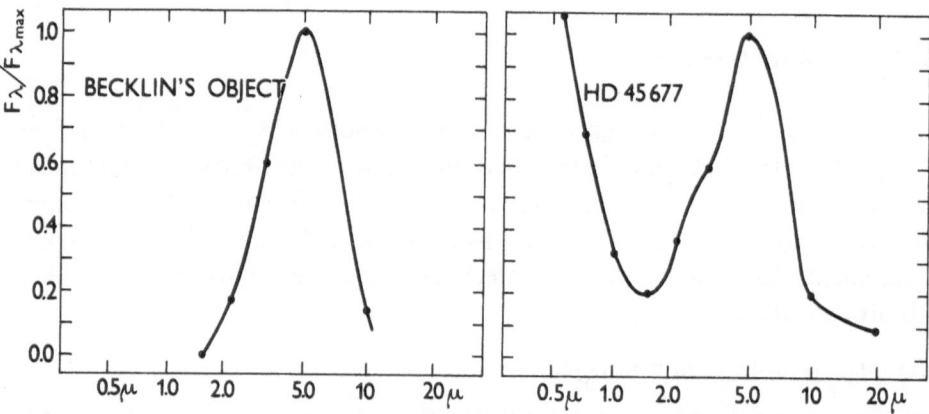

Fig. 2.10.   Infrared stellar sources. Becklin's object is not associated with a visible star. The circumstellar dust cloud is apparently as thick that the central star can not be seen. In the second example a Be star is visible which is embedded in a dust cocoon. The dust is partly transparent to light and transforms UV radiation into IR wavelengths.

black-body. Some IR stars, such as NML Cygni, which are similar to the Becklin object may be either intrinsic IR radiators or highly reddened supergiants. The stars in the third group (e.g. HD45677 – a Be star; T Tauri stars; R Monocerotis, etc) are strong IR sources and at the same time visible in light. The IR excess in their spectra is apparently due to circumstellar dust clouds. Fluxes of light and UV radiation from the central star are absorbed and reradiated in the IR by the circumstellar dust particles. The transformation may be total (Becklin's object) or partial (as for the star HD 45677). The circumstellar dust may be either a remnant of the protostellar material or it may be a product of violent ejections from the central star itself. The genesis of the stars surrounded by massive dust clouds is not yet clear.

It has been suggested that some bright IR stars may be projectors surrounded by protoplanetary clouds (Section 5.5). They would represent early evolutionary phases of planetary systems. The Protosun enveloped within a dust cloud formed by crushing all the planets might have been similar to an IR star such as R Monocerotis.

(d) *Infrared Nebulae*

Hot clouds of interstellar plasma, i.e. emission nebulae, are also sources of IR radiation. There are two classes of such nebulae, viz. planetary nebulae and bright diffuse nebulae. The planetary nebulae are comparatively small, shell-like structures of hot plasma expanding away from the very hot (3×10^4 to 10^5 K) central exciting star. Bright diffuse nebulae are concentrations of hot interstellar plasma, often asymmetrical, surrounding early type stars. In denser plasmas with temperature higher than 4×10^3 K hydrogen is strongly ionized and such plasmas are also called H II regions.

Many planetary nebulae are strong IR sources. Their radiation is probably thermal emission by graphite grains heated to a temperature of about 200 K by the UV photons from the central star. H II regions in general are strong emitters of both near and far IR radiation – in continuum and lines as well. Their luminosities for 45 to 750 μm vary from 10^4 to 10^7 L$_\odot$. Their color temperature is 65 to 120 K and the mechanism of emission in the far IR is explained as reradiation by the heated dust component of the H II regions. The dust is heated by the UV radiation of the embedded or close hot stars.

Besides the thermal emission of dust grains the H II regions and planetary nebulae have IR line spectra of H, He, C, Ne, and S.

(e) *Infrared Galaxies*

A significant fraction of extragalactic objects are strong IR sources, for example: the nuclei of normal galaxies (as in our Galaxy), exploding galaxies (M 82), Seyfert galaxies (NGC 1068), N-galaxies (3C 120), and quasars (3C 273). Infrared luminosities of galaxies range from 10^{37} to 10^{48} erg s^{-1}. The number of IR galaxies is relatively large which means that the strong IR emission is released during a considerable fraction of the lifetime of a galaxy.

Time variations of the order of day have been recorded on different wavelengths (2 μm, 5 μm, 10 μm, 22 μm) which limits the size of the emitting region to about 10^{15} cm or less. The mass producing the radiation must therefore be relatively small. The luminosity of galactic nuclei (like 10^{46} erg s^{-1} from 3C 120) must be emitted for a considerable fraction of the galaxy lifetime, since otherwise such galaxies would be more rare than observed. The duration of the enhanced IR luminosity is estimated to be 10^8 to 10^9 yr, i.c. 10^{15} to 10^{16} s. The total energy output in the form of IR radiation should be about 10^{62} erg, which corresponds to a rest energy of 10^8 $M_\odot c^2$. This in turn means that a considerable fraction of the rest energy in the central parts of a galaxy must be released.

It is not yet fully understood by what mechanism the energy is released. Thermonuclear reactions seem insufficient to explain the immense energy output (for 3C 273, 10^4 times more than the total light luminosity of our Galaxy). Gravitation seems to be the only force which is more effective than thermonuclear reactions (if annihilation of matter and antimatter is not considered). Thus gravitational contraction or collapse is the probable energy source for IR galaxies.

(f) *Infrared Background*

Well over 2×10^4 IR sources have been registered. They are superposed on the IR background. The background consists of atmospheric, interstellar and extragalactic components. Our atmosphere radiates IR lines of O I as rocket experiments have shown. Some ions in interstellar space (e.g. C II at 156 μm, Si II at 35 μm, and Ne II at 13 μm) have been considered as possible sources of monochromatic IR background radiation.

The main source of the IR background must be nuclei of galaxies. Luminous young galaxies formed at large distances (with redshifts larger than 2) may represent a substantial contribution to the IR background, due to the redshift of their light photons to the IR.

The IR background has been measured by several groups of scientists from the ground, from balloons, rockets, and space vehicles. The measurements are difficult and one often receives only upper limits to the fluxes.

2.2.6. VISIBLE UNIVERSE

The paramount importance of light photons (5×10^{14} Hz $< \nu < 10^{15}$ Hz, 0.4 μm $< \lambda < 0.7$ μm) for our lives is due to the sensitivity of our retina which corresponds to the spectral distribution in the solar radiation. This sensitivity is a result of the evolutionary adaptation of our eye to the energy distribution in the solar spectrum.

During daytime our light environment consists of photons of solar origin. They are scattered, reflected, refracted, absorbed, diffused or otherwise transformed with respect to their direction, intensity, wavelength, or polarization. If they are focussed on our retinas they yield information about shape, color, distance, position, motion, etc.

The light photons incident on green plants is partly transformed by chlorophyl into chemical energy of hydrocarbons, fats, and proteins. By this photochemical process, called photosynthesis, a constant flow of solar energy is supplied to the biosphere. The light photons as energy carriers sustain all life on the Earth, including our own.

Our photon environment during nighttime is partly man-made and partly natural. The artificial photon environment derives its energy also from solar radiation (fire, electricity from fossil fuels, hydroelectric power). By illumination of streets and buildings man pollutes the natural light environment. Astronomical observatories, built outside cities a hundred years ago, have been swallowed by modern suburbs and some of the photons entering the telescopes are man-made, i.e. light of street lamps scattered by the atmosphere.

The natural light of the night sky has several components: airglow (photochemical excitation of the upper terrestrial atmosphere), aurora (excitation of the upper atmosphere by energetic particles), zodiacal light (sunlight scattered by interplanetary dust), diffuse galactic light (starlight scattered by interstellar dust), integrated starlight (direct light of stars of our Galaxy) and isotropic extragalactic light. On occasion the Moon and planets reflect sunlight towards the dark hemisphere of our planet.

(a) *Light Background*

It is a difficult task to ferret out the isotropic background from the light of stars of our Galaxy, from the airglow of our atmosphere, and from the solar light scattered by planetary dust. Upper limits of its intensity can be determined only in a few narrow wavelength intervals (e.g. near 5500 Å and 4100 Å). There are sound theoretical estimates of the isotropic light background, which depend upon the cosmological redshift and upon the evolution of galaxies, which are the main contributors.

(b) *Light Sources*

Individual sources of light photons are superposed on the diffuse light background of the sky: the Sun against the blue light scattered by atmospheric molecules; the Moon, planets, satellites, meteors, comets, nebulae, and galaxies upon the unpolluted night sky. Some of the luminous bodies are selfluminous, producing photons from their own energy resources: stars from thermonuclear reactions; protostars, supernovae, IR galaxies, exploding galaxies, quasars from gravitational energy; meteors and supernova remnants from kinetic energy. Other luminous bodies are illuminated (satellites, planets, comets, zodiacal light, and asteroids) so that their light consists of solar photons which are reflected, scattered, diffused, or otherwise transformed. Also, light from concentrations of cold interstellar dust and gas (i.e. from reflection nebulae) is the reflected radiation of nearby stars.

Spectral decomposition of the light from luminous bodies has been until now the main source of information about their nature and properties. Chemical composition, temperature, pressure, magnetic fields, gravitational field, size, distance and motion of the observed source are deduced by applying basic physics to lines, bands and continuum in its spectrum. Our knowledge of the structure and evolution of the universe has been achieved by collecting photons from luminous sources at the focus of telescopes and by sorting the photons in spectrographs according to their energy. It is surprising that astronomers have learned so much from the photons of light, which cover only one of the fifty octaves of the electromagnetic spectrum.

Not all objects may be seen in light photons. Some X-ray, IR, and radio sources have not been identified as optical sources. The luminous universe is only part of the universe observed in the whole range of the electromagnetic radiation. But even the entire photon-emitting universe does not represent the complete universe. Collapsed material bodies with dimensions smaller than their gravitational radius (see Section 2.4) can never emit electromagnetic radiation of any kind.

2.2.7. ULTRAVIOLET SKY

The UV photons (100 Å $< \lambda <$ 4000 Å; 10^{15} Hz $< \nu < 3 \times 10^{16}$ Hz) are absorbed by the Earth's atmosphere and by photoionization of the interstellar H I for $\lambda < 912$ Å. The universe is observable at $\lambda > 912$ Å from rockets and satellites. However, for the extreme UV, i.e., for $\lambda < 912$ Å, the interstellar space is opaque (Figure 2.11). The

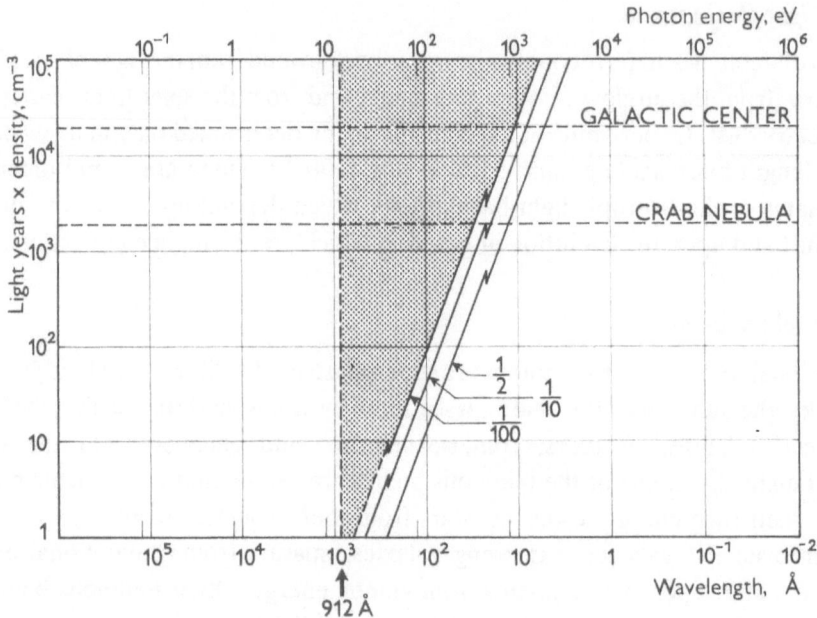

Fig. 2.11.   Absorption of UV and X radiation by the interstellar gas. The absorption edge at 912 Å corresponds to ionization of the interstellar H I.

photons of the extreme UV have energy >13.6 eV and ionize H I from its lowest energy state. Hydrogen atoms are the most abundant in the interstellar medium and hence its high opacity. However, the probability of photoionization decreases rapidly with the photon's energy (photoionization cross section $\sigma \propto \nu^{-3}$), so that from the nearest stars photons with $\lambda < 100$ Å ($h\nu > 100$ eV) can penetrate to our Earth; the $\lambda \sim 50$ Å photons reach us from distances 10^2 pc and even more; for $\lambda < 50$ Å the galactic disk becomes transparent; distant galactic objects may be observed in photons with $\lambda < 10$ Å ($h\nu > 1$ keV). But the photons with $\lambda < 100$ Å are already considered X-ray photons. Direct observations of the isotropic cosmic background are therefore impossible for 100 Å $< \lambda < 912$ Å.

(a) Ultraviolet Background

The visibility in the extreme UV depends on wavelength and also on direction because the interstellar H I is not isotropically distributed. Indirect intensity estimates for the isotropic background radiation can nevertheless be made by observing at λ 21 cm the distribution of the H I at the periphery of other galaxies, because H I there absorbs the isotropic UV radiation and becomes ionized. The H I distribution has been measured in tens of galaxies and an upper limit $I < 10^{-23}$ erg cm^{-2} s^{-1} sr^{-1} Hz^{-1} for the extreme UV isotropic background has been deduced.

The UV radiation $\lambda > 912$ Å from extragalactic sources can penetrate into our planetary system and is measured by instruments outside the Earth's atmosphere.

(b) *Ultraviolet Planets*

Ultraviolet spectrophotometry is a powerful tool for exploration of the solar system. Many atomic and molecular species have strong emissions or absorptions in their UV spectrum. The spectrum therefore provides information about atmospheric constituents, excitation processes and energy equilibrium in the upper part of planetary atmospheres.

Venus shows strong atomic lines, such as: H_I 1216 Å (resulting from scattering of the solar $L\alpha$ in the upper atmosphere of the planet); O_I 1304 Å and 1356 Å (which show that there is an important amount of oxygen in the upper atmosphere of Venus: about 10% as much as the most abundant constituent, CO_2). In the Martian atmosphere, the ratio of O_2 to CO_2 is roughly about 1%.

The atmosphere of our Earth is also the source of UV radiation. In the transition region between the atmosphere and interplanetary space (i.e. the exosphere and geocorona extending to $\sim 15\ R_E$) H and He are the main constituents. The lines $L\alpha$ 1216 Å and $L\beta$ 1026 Å from H_I, and 584 Å from He_I atoms have been observed.

Observations of comets reveal an extensive $L\alpha$ halo. The H_I in the halo is a dissociation product of water sublimated from the cometary nucleus. In studied comets thus far the halo contains about 10^{12} g of H. The H halo is optically thick for $L\alpha$ radiation to a distance of $\sim 10^5$ km from the nucleus.

(c) *Ultraviolet Sun*

The UV spectrum of the Sun has several interesting features. Between 2000 and 1500 Å Fraunhofer absorption lines fade out and emission lines become prominent. At $\lambda < 1500$ Å the continuum is free of Fraunhofer lines and $L\alpha$ is the strongest emission line of the spectrum. $L\beta$ and more than ten members of the Lyman H series are distinguishable before the limit 912 Å is reached.

Most of the lines in the solar UV spectrum are either resonance lines or lines with a low excitation potential of different ionization stages of the abundant elements (C_{IV}, N_{III}, N_V, O_I–O_{VI}, Ne_{VIII}, Mg_X, Si_{XII} etc.). These various lines originate at different levels in the chromosphere and in the transition layer between the chromosphere and corona.

(d) *Ultraviolet Stars*

The bulk of the radiation of early-type stars (e.g. Sirius, Regulus, Bellatrix) lies in the UV. The luminosity of these stars is substantially greater than that of our Sun. Many absorption lines have been found in their continua. The UV spectra of stars are strongly influenced by fine interstellar dust (UV extinction) and by atoms of interstellar gases. The strength of the H_I absorption line $L\alpha$ is in agreement with H_I densities derived from λ 21 cm data in some parts of the sky (in the Scorpius region), but there is disagreement between $L\alpha$ and λ 21 cm data in some other parts of the Milky Way.

Interstellar Lα has been observed from the spacecraft Venera and Mariner. There are several contributions to the Lα diffuse radiation: the interplanetary component, the galactic component which is probably mainly due to circumsolar H with a radius of about 50 pc and partly also to the Lα radiation of the H I regions.

(e) *Ultraviolet Galaxies*

Galaxies are also sources of UV radiation. The central part of the Andromeda Galaxy (M 31) shows a strong excess of UV emission. This may be due to a high proportion of hot stars. It is evident that the hotter a star is the more it radiates in the UV. Hot stars are important contributors to the integrated radiation of galaxies.

Young galaxies must have been strong sources of UV radiation. In the collapse of the Protogalaxy before star formation took place, about 10^{59} erg were released from gravitational energy. A large fraction of the released energy was in the form of Lα photons. As the contraction might have lasted for 10^{14} to 10^{15} s, the UV luminosity of the Protogalaxy was 10^{44} to 10^{45} erg s^{-1}. This is comparable to the present total luminosity of our Galaxy. If the thermonuclear sources are included and if in the beginning early-type stars were formed from the protogalactic gas, then the output of UV photons would have been substantially higher.

The young galaxies that were born very early (redshift z 20 to 30) contribute to the IR isotropic background, but not to the UV background. On the other hand, if galaxy formation continued to later periods (to redshift $z \leqslant 1$), the radiation of such young galaxies would fall principally in the UV spectral region.

The intergalactic gas is often considered as a possible source of UV radiation (i.e., its isotropic background). The observational radio data (λ 21 cm) and measurements of soft X-radiation (44 to 60 Å) indicate that the intergalactic plasma should have a temperature in the interval 10^4 to 3×10^6 K, if the mean density of the matter in the universe is higher than the critical density ϱ_{cr} (10^{-29} g cm^{-3}). If the intergalactic plasma has such a density and its temperature is in this range, it should radiate principally in the UV. Therefore the measurements of the UV background at $\lambda > 912$ Å should give information about the intergalactic gas and mean density of the universe. The value of the mean density is decisive for the future evolution of the universe. If it exceeds the critical density ϱ_{cr} then the universe is closed and the present expansion will eventually stop and reverse into contraction.

2.2.8. X-RAY SKY

Fluxes of X-ray photons with energies ranging from 100 eV to 1 MeV (i.e., 0.01 Å $<$ $\lambda < 100$ Å) are incident upon the Earth. The terrestrial atmosphere is, however, a strong absorber of X-radiation (Figure 2.3) so that observations of the X-ray sky have to be made from above the terrestrial atmosphere. The appearance of the X-ray sky shows some similarity to the radio and optical sky (Figure 2.12). The Milky Way is apparent, as is the concentration towards the galactic center. The number of sources decreases with increasing latitude and it is possible to observe extragalactic sources in high latitudes.

The X-ray sky has a diffuse background component on which individual discrete sources are superposed. The most complete list of X-ray sources – the 3 U catalogue – contains 161 discrete sources. It covers about 85% of the sky in the region 2 to 6 keV, to a sensitivity 2×10^{-10} erg cm^{-2} s^{-1}. The catalogue represents results of measurements from the satellite UHURU, launched on 12 December 1970. From the third UHURU catalogue, about 100 lie in our Galaxy (Figure 2.12).

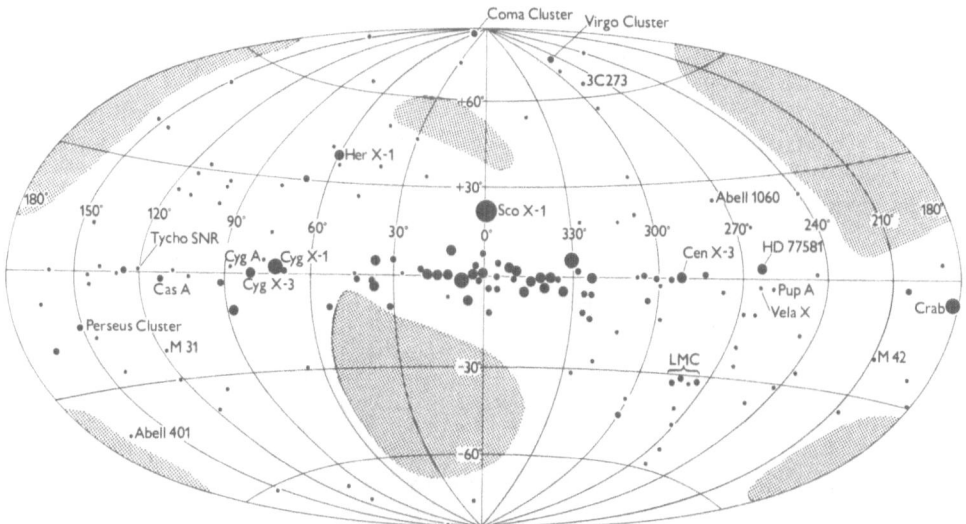

Fig. 2.12. X-ray sources as seen by the UHURU satellite. The grid superposed on the map is in galactic coordinates. The dots show approximately position of the sources and their size corresponds to their brightness. Different classes of X-ray sources may be seen: pulsating X-ray sources (Cen X-3, Hercules X-1), black holes (Cyg X-1), supernova remnants (Crab, Vela), normal galaxies (M-31, Large Magellanic Cloud), radio galaxies (Cygnus A), quasars (3 C 273), clusters of galaxies (Coma cluster). (UHURU catalogue).

Some of the X-ray sources have been identified with optical objects, making it possible to distinguish their distances as well as the X-ray luminosity of the sources. The galactic X-ray sources are either compact X-ray sources (also called X-ray stars) or supernova remnants (Figure 2.13). The X-ray stars are very luminous objects, so that it has been possible to identify some in other galaxies.

The X-ray emission of normal galaxies is integrated emission from the individual X-ray sources. Certain special properties of a galaxy make it a strong X-ray emitter. Peculiar galaxies (Seyfert galaxies, N-type galaxies, exploding radio galaxies, and quasars) have X-ray luminosities several orders of magnitude greater than normal galaxies (such as our Galaxy). It seems that the X-ray emission from clusters of galaxies originates in their intergalactic space.

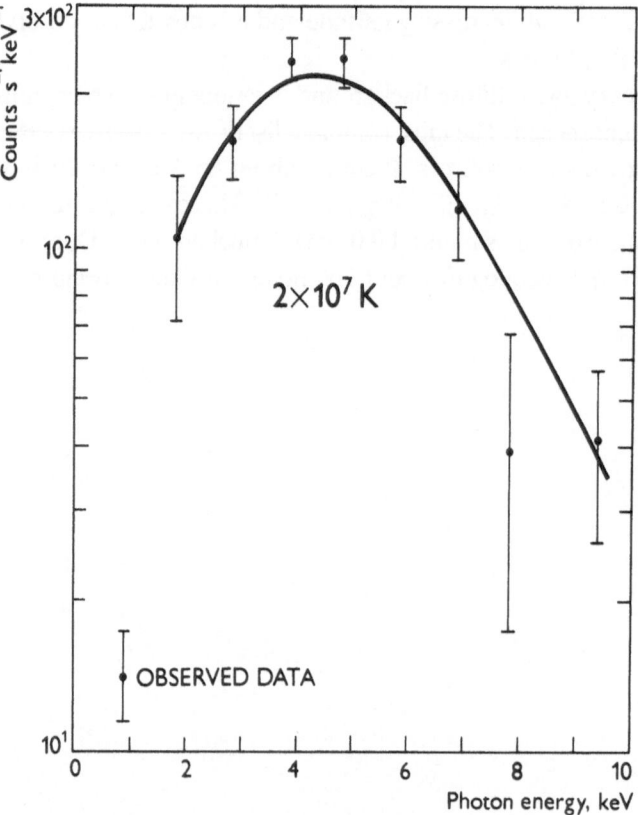

Fig. 2.13.   Spectrum of a compact X-ray source Cyg X-3 may be interpreted as a black body radiation of
2×10^7 K (Giacconi).

(a) *X-Ray Background*

The isotropic background in the spectral region 1 keV to 1 MeV has been studied by many observers and theoreticians. In the energy range 1 keV to 100 keV the radiation is remarkably isotropic but below 1 keV the isotropy stops. The spectrum of the background may be expressed by two power functions viz. with exponent $\gamma = 1.7$ to 1.8 for photons up to 40 keV and $\gamma = 2.2$ to 2.4 for photons with energies 40 keV to 1 MeV.

The origin of the X-ray background is not yet completely understood. It may be emitted by many distant and unresolved extragalactic sources. Some of the background radiation may derive from the fossil microwave photons which are very numerous in cosmic space. The inverse Compton effect transfers energy from relativistic electrons to the fossil photons, which change into X-ray photons.

For the soft X-rays (0.15 to 1 keV) the situation is rather confusing, since accurate measurements are hampered by the low efficiency and poor energy resolution of the existing X-ray detectors. Absorption by interstellar gas should reduce the isotropic component at low galactic latitudes but no conspicuous decrease of the intensity near the galactic plane has been observed.

(b) X-Ray Aurora

High energy photons are generated by electromagnetic interactions in the Earth's atmosphere. One component of the secondary cosmic radiation generated in the terrestrial atmosphere contains electrons and energetic γ-photons with energies >1 MeV. Precipitation of electrons in aurorae (with energies ~ 10 keV) produces X-ray photons above 90 km, which then propagate downwards. Auroral X-ray photons have been recorded from balloons in geomagnetic conjugate points in both hemispheres simultaneously. It seems that the X-ray aurora can be detected over a much larger area than visual aurora. Simultaneous X-ray flashes observed in both hemispheres are used to detect instantaneous conjugate points and their motion.

(c) X-Ray Sun

The first cosmic source observed in X-rays was the Sun. A rich observational material has been accumulated on fluxes, spectra, distribution of X-ray intensity on the disk of the quiet and active Sun.

The X-ray flux is very sensitive to solar activity. It varies within wide limits, from 10^{-4} erg cm^{-2} s^{-1} when the Sun is quiet to several erg cm^{-2} s^{-1} (for photons with energies from 0.5 to 10 keV). The Sun remains always at least three orders of magnitude brighter than the brightest extrasolar X-ray source, Sco X-1, the flux of which is 100 keV cm^{-2} s^{-1}.

Most of the solar X-ray photons are emitted from coronal condensations above active regions. The strong local magnetic fields in the solar photosphere produce sunspots, plages in the chromosphere, and coronal condensations in the overlying corona. In the condensations the density and temperature are considerably increased when compared with the corona outside active regions. The hot and dense plasma is bound to magnetic tubes of force which have the form of loops, arches, or straight columns with diameters of a few arc seconds (one arc second from the Earth means about 725 km on the Sun). Thus the structure of coronal condensations, i.e. X-ray sources, is highly inhomogeneous in density, temperature, and magnetic fields. Strong currents are supposed to flow along the tubes. The X-ray photographs of the Sun indicate structure of magnetic fields protruding from the photosphere (or sunspots) high into the coronal and eventually into the interplanetary space. Bright X-ray emission features match closely the geometry and intensity of the coronal magnetic fields in active regions.

In addition to a few active regions on the Sun, many small ($\sim 10\,000$ km) bright X-ray points exist on the Sun. About one hundred of them are visible on a typical X-ray photograph. The bright points represent small, low-lying and closed bipolar magnetic fields. They probably represent emerging magnetic fluxes. As such they would be an important contributor to the solar magnetic field.

The X-ray spectrum consists of recombination continua of highly ionized ions and of their lines. Among the ions one finds O VII, O VIII, Fe XV to Fe XXV, Mg X to Mg XII, Si XIII, Si XIV, etc. The resonance lines of all the highly ionized ions lie in the spectral range $\lambda < 25$ Å. There is also a substantial continuum emission.

The X-rays are of great importance for solar physics because they carry information about the chemical composition of the coronal plasma, about the structure of
coronal condensations and their corresponding magnetic fields, about the state of the
solar corona, and also about the processes of X-ray emission from the hot coronal
plasma. The Sun is the only cosmic X-ray source in which spectral lines have been
measured spectroscopically in a wide range of densities (10^{-16} to 10^{-14} g cm^{-3}) and
temperatures ($10^6 < T < 10^8$ K).

The solar X-ray emission has an important role for solar-terrestrial relations
because it ionizes upper layers of the Earth's atmosphere (ionosphere). The ionosphere shields the terrestrial biosphere against lethal effects of the solar X-ray
photons and it is important for radio communication.

(d) X-Ray Stars

This type of galactic sources is also called 'compact X-ray sources' or 'stellar X-ray
sources'. About 50 of them have been detected.

The Sun is a normal star and many other stars with a hot corona probably also emit
X-ray photons. But normal stars are inefficient X-ray emitters (for our Sun the X-ray
luminosity is six orders of magnitude smaller than its optical luminosity). Due to their
large distances, the X-rays of normal stars are not detectable by present observing
techniques. In the X-ray stars the relation of both luminosities is reversed: they have
a maximum luminosity of 10^{38} erg s^{-1} and minimum luminosity of 10^{36} erg s^{-1}. That
means that the X-ray luminosity is about ten orders of magnitude larger for the
X-stars than for our Sun and other normal stars. Special circumstances are needed
for a star to be an X-ray source.

The source Sco X-1 with its flux of 100 keV cm^{-2} s^{-1} is the brightest X-ray source
in the sky (besides the Sun). Its discovery in 1962 marked the beginning of stellar
X-ray astronomy. It has the unique distinction of being observable in the X-ray, UV,
visible, IR, and radio spectral ranges. Its X-ray spectrum has an exponential
character, $I_\nu \propto e^{-h\nu/kT}$, which suggests thermal bremsstrahlung. Fitting an exponential curve to the spectral distribution, we obtain a temperature value, $T = 5 \times 10^7$ K.
At such temperatures Sco X-1 should be a plasmatic body. The spectrum in the
optical and IR regions is flat, indicating that the emitting plasma is optically thin. At
very low frequencies the plasma should again be self-absorbing, with $I_\nu \propto \nu^2$.

The optical spectrum of Sco X-1 (it appears as a star of magnitude 13) shows a
similarity to spectra of old novae. It is known that such objects are close binaries in
which mass is transferred from one member to a white dwarf component. For some
X-ray stars the binary nature has been confirmed by observations (e.g. Her X-1 and
Cen X-3, Cyg X-1 and Cyg X-3). Such observations support the idea that the X-ray
stars are close binary stars, in which the primary component (an ordinary star)
transfers matter to a compact companion. Matter can be transferred from a massive
primary by stellar winds, or a luminous X-ray source may heat the outer layers of the
primary to produce a wind. There is a third possibility of mass transfer from the
primary to the compact companion, viz. the expanding primary may exceed the size

of the critical Roche surface and its plasma is accreted by the compact companion. The gases falling towards the companion gain kinetic energy and produce a hot turbulent plasma region near the companion's surface. If the mass of the companion is M and its radius R, then by thermalization

$$kT \sim \alpha G m_p M/R, \qquad (2.12)$$

where α is an efficiency factor, depending on the mechanism of heating. It may have values from 0.1 (adiabatic heating in shock waves) to 10^{-6} (viscous heating in a disk around the companion). Expressing mass and radius in solar units $M_\odot$ and $R_\odot$, the relation (2.12) is

$$T \sim 10^7 \alpha (M/M_\odot)(R_\odot/R) \text{ K}. \qquad (2.13)$$

If the accretion has to produce temperatures of about 10^8 K, the radius R must be small. The companion must therefore be either a white dwarf, a neutron star, or a black hole. A rapidly fluctuating X-ray source Cyg X-1 was also optically identified with a peculiar binary system in which the invisible companion has a mass of the order of 10 solar masses. Such a mass cannot belong to a white dwarf or a neutron star (see Section 4.5.4); it should be a black hole.

(e) *Supernova Remnants*

The other group of galactic X-ray sources are supernova remnants. Seven supernova remnants have been detected among the X-ray sources. The young supernova remnants (less than 10^3 years old) have diameters of a few parsecs and hard X-ray spectra. The old remnants with ages measured in thousands of years (the Cygnus Loop is more than 10^4 yr old), have soft X-ray spectra and X-ray diameters measured in the tens of parsecs. The Vela supernova remnants which are 20 000 yr old have an X-ray diameter of 40 pc.

The Crab Nebula is the best known supernova remnant and a well studied X-ray source. It is the remnant of a historical supernova explosion (July 4, 1054). Its X-ray flux of 15 keV cm^{-2} s^{-1} in the range 2 to 10 keV comes from an area equal to the optical nebula. The spectral distribution follows a power law and is evidently of a quite different origin than is the X-ray star Sco X-1. It is synchrotron radiation emitted by relativistic electrons in the magnetic fields of the nebula.

The Crab Nebula and Vela remnants are the two known examples of supernova remnants with an X-ray pulsar. About one tenth of the X-ray radiation of the Tau X-1 source comes from the pulsar designated NP 0531. Huge quantities of kinetic energy are stored in the rapid rotation of the pulsar. Its period of rotation is increasing with a loss of rotational energy of about 10^{38} erg s^{-1}, which agrees with the luminosity of the pulsar and the remnants. This agreement of both values leads to the conclusion that the rotational energy of the neutron star (pulsar NP 0531) is fed into the nebula and radiated away. The fast rotation (30 times in a second) and intense magnetic fields of the neutron star (see Section 4.5.3) produce large scale electromagnetic fields, which accelerate electrons to relativistic energies ($\geqslant 1$ MeV).

In the absence of such a source, the nebula would have stopped radiating still during medieval ages, only one or two centuries after the supernova explosion.

It seems that most of the supernova remnants draw energy from another source, viz. the thermal energy of the hot plasma which is produced by interaction of the supernova remnants with the interstellar medium. The kinetic energy of the expanding supernova remnants is sufficient for the X-ray luminosity of 10^{36} erg s^{-1} for 10^4 yr or more.

(f) Extragalactic Sources

Integrating the luminosities of individual galactic X-ray sources, one can estimate the total X-ray luminosity of our Galaxy. It is about 10^{40} erg s^{-1}. This is 10^4 times less than the optical luminosity of our Galaxy, so that it is an inefficient X-ray emitter. The mean density of X-ray photons in the interstellar space is $\sim 5 \times 10^{-5}$ eV cm^{-3}, while the light photon density is about 0.5 eV cm^{-3}.

Normal galaxies are inefficient X-ray sources. Both the Magellanic Clouds and Andromeda galaxy M 31 are also X-ray sources. The Small Magellanic Cloud contains a very luminous stellar X-ray source, which is a binary star. Its X-ray luminosity is $L_x = 1.3 \times 10^{38}$ erg s^{-1}. The luminosity $L_x = 3.8 \times 10^{38}$ erg s^{-1} (2 to 7 keV) of the Large Magellanic Cloud is also dominated by four discrete sources. The ratio L_x/L_v for normal galaxies is small, about 10^{-4} or so.

The UHURU satellite has revealed more than 60 X-ray sources in higher galactic latitudes (i.e., not shielded by the absorbing layer of the galactic hydrogen). About half of them have been identified with extragalactic sources: with ordinary galaxies (such as M 31 or the Magellanic Clouds), with radio galaxies (such as Cyg A, Cen A), Seyfert galaxies (e.g. NGC 4151), clusters of galaxies (such as the Virgo cluster, and Coma cluster) and quasars (e.g., 3C 273).

There is a clear distinction between the X-ray emission from individual extragalactic objects and clusters of galaxies. The cut-off in the low energy photons from individual galaxies indicates that the radiation had to pass from a small central region through the absorbing interstellar matter of the galaxies. The cluster sources are extended and have no cut-off. Their X-ray radiation is emitted by extended intergalactic spaces of the clusters – it is not integrated radiation of individual members.

2.2.9. GAMMA SKY

Thanks to modern technology, observations of γ-photons from the universe are possible. The γ-region of the electromagnetic spectrum (above 1 MeV) is a natural continuation of the hard X-ray region and even some mechanisms of emission are common to both, e.g. bremsstrahlung and inverse Compton effect. Other mechanisms are specific for γ-photons: (a) electromagnetic decay of π^0 and Σ^0

$(\pi^0 \to 2\gamma, \Sigma^0 \to \Lambda + \gamma)$; (b) annihilation $(e^+ + e^- \to 2\gamma)$; (c) deexcitation of nuclei $(n + p \to {}^2H + \gamma)$.

Observations of the γ sky are difficult, because the number of incident photons is very low when compared with the background of cosmic rays. Gamma photons are routinely produced in terrestrial accelerators and the techniques for their detection are used for observing the γ sky.

In our planetary system the natural γ radiation has been measured from three bodies. On our Earth, some elements of the lithosphere are γ radioactive and a steady flux of γ-rays is produced in the atmosphere by high energy cosmic particles. On the Moon, the natural and cosmic-ray induced γ activity has served for the determination of the chemical composition of the lunar surface. Gamma rays from solar proton flares were detected for the first time in 1973, August 4 to 7.

The protons accelerated during a flare interact with the the nuclei of C, N, and O in the solar atmosphere. The interactions result in a spallation reaction or the protons are inelastically scattered, leaving part of their kinetic energy to the target nucleus which becomes excited. The excited nuclei then deexcite by emission of γ-photons. Neutrons knocked out by spallation reactions are partially captured by protons to produce deuterons and 2.23-MeV photons. The emission line corresponding to this energy has been recorded during the afore-mentioned flares and its intensity is proportional to the flux of neutrons in the flares.

Another emission line observed in solar flares corresponds to positron-electron annihilation (1.15). The positrons are produced by decay of π^+ mesons (2.6) and β^+ decay of spallation products.

Although the observational data for the γ sky are still scarce and sometimes rather uncertain, some discrete γ-sources have been found. The central regions of our Galaxy are a source of γ-radiation, although it is not known whether the source is diffuse or discrete. Its γ-flux has been found to be about four times greater than the flux from other directions.

Gamma photons have been recorded from all other directions in the sky, but not with the same intensity. There is a pronounced concentration towards the galactic plane and especially towards the galactic center (Figure 2.14). There are some theoretical calculations of the γ-background, but more experimental data are needed to find its true nature.

Enhanced γ-radiation (γ-sources) has been observed in several directions in addition to the center of the Galaxy. The Crab nebula spectrum has been measured up to 100 MeV. The spectrum may be represented by a power function $\alpha E^{-\gamma}$ with γ equal 2.2 to 2.4. Some sources may be variable. Thus the source found in Libra, which appeared clearly in one balloon flight, did not appear during a flight 9 mo earlier. A number of soft γ-bursts have been detected by Vela satellites and the direction of arrival could be determined. A large number of different theoretical explanations then appeared concerning the nature of the source. Only more solid observational data will decide which of the theories, if any, corresponds to reality.

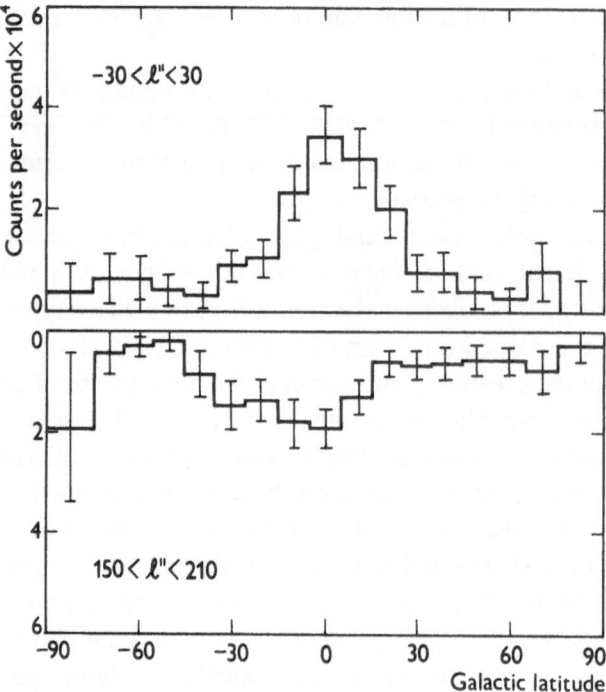

Fig. 2.14.   Gamma-ray intensity and its distribution in our Galaxy. The upper curve gives the distribution in latitude in the direction of the galactic center. The lower curve is for opposite direction towards the anticenter. The gamma-ray increase towards the galactic center is obvious.

2.2.10. Large scale electromagnetic interactions

The electric charge of individual particles does not usually act over large distances, because there is a strong tendency for positive and negative charges to be balanced everywhere. Nevertheless, macroscopic electrostatic fields are also encountered in the universe.

(a) *Electrostatic Fields*

An electrostatic potential of about 380 000 V exists between the ionosphere and the Earth. It is sustained by storms all over the globe. We live in an electrostatic field which is locally influenced by storms and shows also some dependence on the solar activity.

   The solid surfaces of the Moon and of interplanetary and interstellar dust particles are being charged by photoemission and by secondary electron emission. The contact with the solar wind or interstellar plasma complicates the situation. It may be shown, for example, that at the subsolar point the lunar surface should be positively charged to a few volts, while at the lunar terminator and dark-side surface the charge should be negative (tens of volts). The charging of the lunar regolith by photoemission, by solar wind bombardment and by particles of the magnetotail of the Earth has two consequences: (a) a tenuous electron-gas layer ($\sim$100 electrons in cm^3) above the sunlit hemisphere is formed mainly by photoemission and (b) electrostatic transportation of the lunar surface material is important for shaping surface features up to a size of a few meters. Charged dust grains are levitated electrostatically a few centimeters above the lunar surface. Forward scattering of sunlight on the grains results in the horizon-glow observed by Surveyor-7 one hour after the local sunset.

The interstellar grains are probably also electrically charged under the influence of UV stellar radiation. The grain material seems to be a mixture of different constituents (graphite, silicate, Fe, quartz, silicon carbide etc.) and the individual components will be charged to potentials of opposite sign. Some grains may be charged by photoelectric emission caused by UV photons. Other particles carry a negative charge due to the capture of free electrons from the surrounding plasma. The growth of composite grains will then be favored by electrostatic attraction. The process of grain growth is of great importance for processes in interstellar space, among others for the condensation of stars and planets.

(b) *Magnetic Field in Plasmas*

Electrostatic fields are due to the separation of positively and negatively charged particles ($\nabla \cdot \mathbf{E} = 4\pi\varrho$, where $\mathbf{E}$ is the vector of the electric field and ϱ is the charge density). Electric fields are induced by changes of magnetic fields ($\nabla \times \mathbf{E} = -(1/c)(\partial\mathbf{B}/\partial t)$ or due to motion $\mathbf{v}$ in a magnetic field ($\mathbf{E} = (\mathbf{v}/c \times \mathbf{B})$. A force $\pm e\mathbf{E}$ acts on a particle with charge $\pm e$ in the electric field $\mathbf{E}$. Moving electric charges, i.e. electric currents, produce magnetic fields in the surrounding space.

Most cosmic bodies in the observable universe (Sun, stars, nebulae, interplanetary matter, interstellar matter and planetary ionospheres) are built up of plasma, either completely or at least partly. Consisting of free electrons and ions, plasma responds by electric currents to any electric field. Electric currents are accompanied by magnetic fields $\mathbf{B}$ and plasma with magnetic fields is called magnetized plasma. Currents in a magnetized plasma result in Lorentz forces $\mathbf{j} \times \mathbf{B}$, where $\mathbf{j}$ is the current density, i.e. the charge flowing through 1 cm^2 s^{-1}. The Lorentz forces in turn change the velocities $\mathbf{v}$ of the moving plasma. The electromagnetic and hydrodynamic phenomena are thus coupled and a special scientific discipline called magnetohydrodynamics, or hydromagnetics, interprets the large-scale electromagnetic interactions in cosmic plasmas. As a result of the coupling, magnetic lines of force are frozen in the plasmas and restrict their motion (see Figure 3.2). Each ion and electron of the magnetized plasma is forced by the Lorentz force to circle around a field line. Hence plasma and the frozen in field share a common fate (Figure 2.15). The fate depends upon which of the two partners in the magnetized plasma is more energetic: the density of kinetic energy or the density of the magnetic field energy, i.e. $\frac{1}{2}\varrho v^2$ or $B^2/8\pi$). Either

$$\tfrac{1}{2}\varrho v^2 > \frac{B^2}{8\pi} \qquad\qquad\qquad (2.14)$$

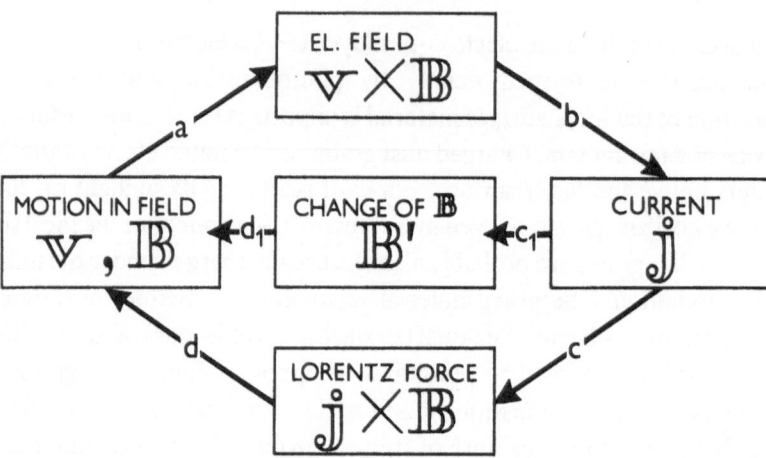

Fig. 2.15. Motion of a plasma in magnetic fields starts a sequence of processes: (a) induction of electric field; (b) electric field in conducting plasma produces currents; (c_1) currents induce additional magnetic fields B' and (c) give rise to a body force on a volume unit of the plasma; (d) the body force (Lorentz force) influences plasma motions, while (d_1) the induced field B' changes the original field B.

or

$$\frac{B^2}{8\pi} > \tfrac{1}{2}\varrho v^2 \tag{2.15}$$

Examples for both inequalities are shown for the case of solar prominences (Figure 3.3) and solar corona (Figure 3.4).

(c) *Magnetosphere*

Magnetic fields are practically everywhere in the universe and they are important for its structure and evolution. Everybody knows that the magnetic field of our planet, the geomagnetic field, governs plasma in the ionosphere, electrons and ions in radiation belts and influences different processes on the Earth's surface, in our atmosphere, and in the circumterrestrial space.

If there were no solar wind, the geomagnetic field would have the symmetric form of a dipole field. But the pressure of the solar wind transforms the geometric field into a comet-like region – called the magnetosphere of the Earth (there are magneto-spheres on other planets – e.g. Jupiter). As a result of deformation by the solar wind, the structure of the magnetosphere is differentiated and, moreover, it is different in the sunwards and in the antisolar directions. Furthermore, the deformation of the magnetic field lines induces different systems of electric currents in the magneto-spheric plasma.

The boundary between the magnetosphere and interplanetary solar wind is well defined on the sunlit side of the Earth. There is a sharp division (magnetopause) between the field lines that are anchored in the Earth and the interplanetary field lines frozen in the solar wind. On the Earth's night side, the boundary is diffuse because field lines of the magnetosphere may be connected to the interplanetary field lines.

The region of closed field lines is called the magnetosphere and its upper boundary is the magnetopause (about $10\ R_E$ in the solar direction). Open field lines in the antisolar direction form the magnetotail, which is about $1000\ R_E$ long. Outside the magnetopause is a turbulent region, called the magnetosheath, where the solar wind is forced by the pressure of the magnetosphere to change its direction and velocity (to subsonic and sub-Alfvénic). The outer boundary of the turbulent region forms a shock wave, where the transition from the ordered solar wind to the very turbulent and disordered magnetosheath occurs. The distance to the shock front in the solar direction is about $14\ R_E$.

The deformation of the magnetic field lines caused by the pressure of the solar wind on the magnetosphere induces currents in the magnetospheric plasma, according to the Maxwell equation

$$\nabla \times \mathbf{B} = \frac{4\pi}{c}\mathbf{j}. \qquad (2.16)$$

There are highly variable magnetic fields (hence the curl of the magnetic field on the left side of Equation (2.16)) just outside the magnetopause, which lead to a system of magnetopause currents. Another current system flows in the neutral sheet which divides the northern and southern polarities in the magnetotail. The field lines of the magnetosphere are bulged with respect to the dipole field, which results in ring currents above the equator. Finally there is a system of currents flowing along the field lines high above the auroral zone.

The magnetosphere protects the terrestrial biosphere from the solar wind and low energy cosmic radiation. Only high energy particles (with energies $> 10^9$ eV) can eventually penetrate the magnetosphere and interact with the atmosphere, where they produce secondary cosmic radiation.

(d) *Moon's Magnetic Fields*

The magnetic field of our satellite is completely different from the terrestrial magnetosphere. While the magnetosphere might be approximated by a magnetic dipole exposed to solar wind, the selenomagnetic field resembles a collection of small magnets buried here and there underground (called magcons which means magnetic concentrations). The maximum recorded selenomagnetic field represents about 1% of the geomagnetic field – i.e., about 400 γ, while the minimum field ~ 5 γ coincides with interplanetary magnetic field intensity. Selenomagnetism is local permanent magnetism, induced about 3.7×10^9 yr ago. It may have been produced by dynamo effect, i.e. by the same mechanism which sustains the geomagnetic field.

(e) *Solar Magnetic Fields*

Many magnetospheric processes would have remained unnoticed had it not been for sensitive detectors aboard satellites. The aurorae in polar regions are an exception.

On the contrary, the hydromagnetic processes (and implicitly electromagnetic interactions) on the Sun, observable from the Earth with modest instruments, belong to the most dramatic spectacles man has ever witnessed. All solar activity (sunspots, plages, flares, prominences, coronal condensations and their structural changes, violent burst observed in radio waves) are manifestations of the interaction of the solar plasma with local magnetic fields.

Due to the finite conductivity of solar plasma, the magnetic field energy $B^2/8\pi$ is partly transformed by ohmic losses into heat. Magnetic field lines frozen in ejected plasma clouds are dragged far away from the Sun, giving rise to interplanetary magnetic fields of a few γ units. The losses of magnetic energy from the Sun are, however, compensated by the supply of kinetic energy from the subphotospheric convective layer and from differential rotation. A field $\mathbf{B}$ (see the left box in Figure 2.15) is increased by convection and turbulence $\mathbf{v}$ by the steps indicated in the Figure: $(\mathbf{v}, \mathbf{B}) \rightarrow (\mathbf{E}' \propto \mathbf{v} \times \mathbf{B}) \rightarrow \mathbf{j} \rightarrow \mathbf{B}'$. This mechanism of transformation of the kinetic energy into magnetic energy continues as long as the inequality (Equation (2.14)) exists in the plasma.

(f) Stellar Magnetic Fields

Magnetic fields exist on many other stars. They are detected only if they are sufficiently strong, so that the splitting of spectral lines by the Zeeman effect is measurable. Peculiar stars of Ap type have fields of 2 to 3 kG and in some cases up to 10 kG. These values represent mean surface fields. There is little doubt that the local fields in some regions of the stellar surface will be substantially stronger. They are thus comparable with iron magnets (up to 10^4 G) and even with superconductive magnets (fields of the order of 10^5 G). The strongest magnetic fields produced in terrestrial laboratories by explosive flux-compression are greater than 10 MG. Still much stronger fields should exist on neutron stars (pulsars), if our understanding of their structure and origin is correct.

During the gravitational collapse of a middleweight star (see Section 5.4) its radius decreases 10^5 to 10^6 times. Hence its surface is reduced 10^{10} to 10^{12} times. The total magnetic flux on the surface should not be changed by the collapse, due to the 'frozen in' lines. The magnetic field on the surface of the collapsed neutron star should therefore be 10^{10} to 10^{12} G, if the original field before the collapse was only 1 G, as is the case of our Sun. In the interior of neutron stars the magnetic field should be still much stronger, $\gg 10^{13}$ G.

The properties of matter in very strong magnetic fields are very different from the properties of normal matter. The structure of the atoms themselves must be drastically changed by the strong external fields, because the Lorentz force on electrons in atoms becomes more important than the internal electromagnetic forces between the electrons and nuclei. For hydrogen this occurs in fields $\mathbf{B} > 2 \times 10^9$ G, because the Larmor radius (Equation (3.5)) of electrons gyrating in the external field becomes smaller than the Bohr radius (see Figure 4.25).

(g) Interstellar Magnetic Fields

The vast spaces between stars are filled with magnetic fields. The interstellar plasma is a good conductor with frozen-in magnetic field lines. But even the clouds of interstellar H I where the abundance of ionized atoms and free electrons is extremely small, are influenced by interstellar magnetic fields. This is a result of their size L rather than of the weak conductivity λ (see Formula 3.8). Some interstellar grains are probably charged and they too should interact with the interstellar magnetic fields.

Since the interstellar medium and interstellar magnetic fields are coupled, the energy densities $B^2/8\pi$ and $\frac{1}{2}\rho v^2$ may be expected to be equal. For $\rho \approx 10^{-24}$ g cm^{-3} and v of the order of 10 km s^{-1} (both values found observationally) the field is about 3×10^{-6} G. The existence of such fields in interstellar space explains various phenomena: the polarization of the light of distant stars, caused by alignment of interstellar dust grains across the field lines and anisotropic absorption; the shape of nebulae elongated in the direction of spiral arms; the Faraday rotation of radiation from extragalactic radio sources; Zeeman splitting of the absorption line $\lambda 21$ cm observed in the spectrum of distant radio sources; and the synchrotron emission on meter wavelengths originating within our Galaxy. There are further arguments for interstellar magnetic fields stretched preferentially along the galactic spiral arms.

The interstellar magnetic fields apparently also participate in the formation of spiral arms, they confine cosmic rays to the Galaxy, and they bind interstellar ions and dust grains (the latter with a charge of 1 to 10 electrons in H I clouds and up to 500 electrons in H II regions) into a system. The fields are important in accelerating ions and charged dust particles to high energies, and they influence the motions of interstellar clouds and shape filaments of nebulae. Being frozen in condensing protostars they represent embryonal magnetic fields of the newly formed star. The kinetic energy of the stellar plasma (in the form of convection, turbulence, differential rotation) then substantially enhances the fossil field inherited from the interstellar space.

The transport of magnetic fields goes also in the opposite direction: stellar winds from magnetic stars, eruptive stars, and exploding supernovae and pulsars which eject magnetic fields with plasma into the interstellar medium. The question whether the interstellar magnetic field pervading the entire Galaxy is primordial or whether it was produced later in the Galaxy remains unanswered. It should be mentioned in this connection, that many other galaxies (and perhaps all) also have interstellar magnetic fields, as is proven by their synchrotron radio emission.

Summary

The electric charge of elementary particles is the source of electrostatic and electromagnetic fields. All the particles with electric charge interact electromagnetically. While strong interactions are of basic importance for nuclei, electromagnetic interactions are important for the structure of atoms, molecules, crystals, minerals, rocks and solid bodies in general.

Accelerated electric charges emit photons, the quanta of electromagnetic fields. The photon energies cover a very broad spectrum, from 10^{-11} to 10^{17} eV. They are emitted from the entire sky and form a diffuse background. Superposed on the relatively weak background are individual discrete sources of photons, such as the Sun, planets, satellites, comets, asteroids, meteors, stars, nebulae, galaxies, radio sources, IR sources, X-ray sources and γ-ray sources. Each source emits photons in one or more spectral ranges. As a result, the appearance of the sky depends on the spectral region used for its survey. Correspondingly one observes the radio sky, IR sky, visible sky, UV sky, X-ray sky or γ-ray sky. Some cosmic objects are observed only in one spectral range, while others are observable in two or more ranges.

Electric charges in motion (electric currents) produce magnetic fields. Magnetic fields are nearly ubiquitous in the observed universe, which consists mainly of plasma – a good electrical conductor. The coupling between plasma and the magnetic fields (via the Lorentz force) plays an important role in many structural features and evolutionay processes of the universe.

2.3. Weak Interaction

There is a clear observational difference between a strong decay of a resonance, e.g. Δ^{++} resonance

$$\Delta^{++} \to p + \pi^+ \tag{2.17}$$

and a weak decay of a Λ^0 particle

$$\Lambda^0 \to p + \pi^-. \tag{2.18}$$

The first reaction occurs so quickly that the parent particle Δ^{++} can never be observed directly. Its lifetime is about 10^{-23} s and even if it moved with velocity close to the velocity of light, its track would be only $\sim 10^{-13}$ cm long and therefore not measurable. On the other hand, the path of the Λ^0 hyperon is easily measurable and its lifetime can be determined directly as 2×10^{-10} s. In both reactions only hadrons are involved and the same type of particles are produced. But the immense difference in the rate of the two reactions – 13 orders of magnitude – indicates that fundamentally different interactions are involved.

2.3.1. WEAK INTERACTIONS

Weak interactions are the weakest of the interactions (except for gravitational forces) between elementary particles. They are responsible for slow decays of elementary particles. While in strong interactions all characteristic numbers are conserved, in weak interactions some conservation laws are violated (strangeness, parity).

There are weak interactions without neutrinos, such as Equation (2.18). But better known and of greater importance in the universe are the weak interactions involving neutrinos. The decay of the neutron is the best known example:

$$n \rightarrow p + e^- + \tilde{\nu}_e. \tag{2.19}$$

The process is known as β-decay, because electrons e^- and positrons e^+ emitted from a nucleus are called β-radiation. Another example of β-decay is

$$p \rightarrow n + e^+ + \nu_e \tag{2.20}$$

which is not spontaneous, because it is endoenergetic. It can occur in a nucleus from which the proton receives the necessary extra energy. For example

$$^{11}C \rightarrow {}^{11}B + e^+ + \nu_e. \tag{2.21}$$

Besides β-emission and β^+-emission (Equations (2.19) and (2.20)), there are other types of interactions characterized by the participation of four fermions:

$$^7Be + e^- \rightarrow {}^7Li + \nu_e \quad \text{(K-capture)} \tag{2.22}$$

and

$$\mu^- \rightarrow e^- + \nu_\mu + \tilde{\nu}_e \quad (\mu \text{ meson decay}). \tag{2.23}$$

The explanation of the weak interactions was given by E. Fermi. His universal interaction theory describes the weak interaction between leptons and baryons by a Hamiltonian which combines the field operators of all four participating particles. All known hadrons and leptons may participate in the weak interaction; it is universal. The weak interaction acts over an extremely short range, less than 10^{-15} cm, while the strong interactions act in the range of $\sim 10^{-13}$ cm.

The effects of weak interactions in the universe are less obvious than those of the other three interactions. They are important for nucleosynthesis (e.g. Equation (2.20) in H transformation into He), as cooling agents in the last phases of thermonuclear evolution of stars (see Section 2.3.11), and the fossil neutrinos or relic neutrinos might represent some missing mass in the universe (see Section 2.3.13).

2.3.2. INTERMEDIATE BOSON

Weak interactions are supposed to be generated through the emission and absorption of a boson, called W-meson or intermediate boson. Its role would be analogous to photons in electromagnetic interactions or pions in strong interactions (see Figure 2.16). To explain weak interactions, the W-meson is supposed to have the following properties: It has an electric charge, i.e. W^+ or W^-. Its mass should be larger than the mass of the K-meson (494 MeV), because if this were not the case the decay

$$(K^\pm \rightarrow W^\pm + \gamma) \tag{2.24}$$

would be observed, which it is not. Theoretical estimates indicate that the mass of the W-meson should be even greater than the proton mass (938 MeV). W-mesons may decay by different modes such as:

$$W^+ \rightarrow \mu^+ + \nu_\mu$$
$$W^+ \rightarrow e^+ + \nu_e$$
$$W^+ \rightarrow \pi^+ + \pi^0 \tag{2.25}$$
$$W^+ \rightarrow \pi_0^+ + K_+^0$$

and similar decays exist for the negative W-meson. Until now, however, the W-meson has not been observed and remains a hypothetical particle.

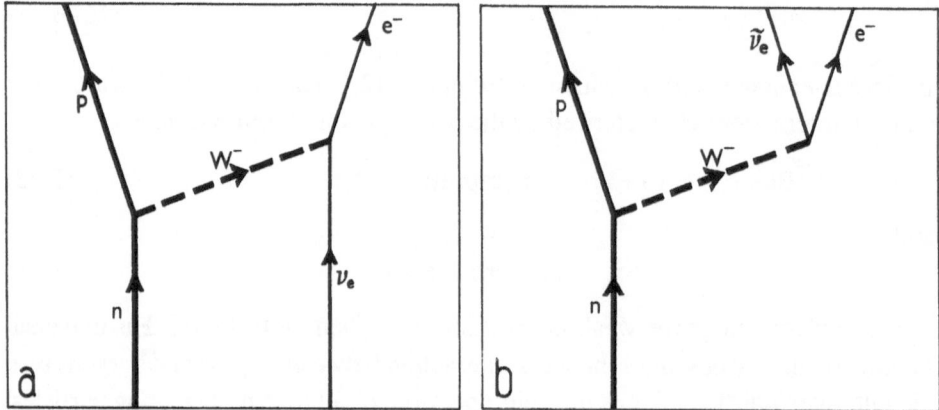

Fig. 2.16. The hypothetical W-boson is supposed to mediate weak interactions such as (a) scattering of neutrinos and (b) decay of neutrons. Time in both diagrams (called Feyman diagrams) proceeds upwards. The W⁻-boson proceeds from left to right – the W⁺-boson would proceed from right to left. The incoming neutrino in (a) is equivalent to the outgoing neutrino in (b).

The large mass of the W-meson follows from the uncertainty principle (the product of uncertainties in position and momentum of a particle, or the product of errors in time and energy determinations cannot be less than Planck's constant). As a consequence, the range of a force should be inversely proportional to the mass of the exchange particles. Electromagnetic and gravitational forces have an infinite range, so that the photon and graviton have zero rest mass. The range of weak interactions ($\sim 10^{-15}$ cm) is shorter than the range of strong interactions ($\sim 10^{-13}$ cm), so that the mass of the W-boson must be larger than that of pions.

2.3.3. NEUTRINO

The existence of neutrinos was proposed in 1931 to account for the missing energy and momentum in β-decays. By a radio-active decay (α, β or γ) the emitting nucleus gets rid of a certain amount of energy corresponding to the energy difference of the nucleus before and after the decay. Emitted α-particles or γ-photons therefore have a monoenergetic spectrum (line spectrum). Also, in a β-decay a fixed amount of energy is liberated from all the nuclei of the radioactive element. But it follows from measurements that the electron or positron transports only a part of the liberated energy, the remainder of the energy being carried away by a neutrino. The β-spectrum is continuous and hence the neutrino spectrum must be continuous also.

The interaction of a neutrino with other particles is very weak and therefore observations of neutrinos are difficult. It is a 'shy', unsociable particle, which escaped direct observations until 1956, when the first neutrino was observed. The advent of fission reactors brought torrential fluxes of antineutrinos. Modern accelerators today are so advanced that they may produce beams of neutrinos for laboratory experiments.

The rest mass of the neutrino is either extremely small or, more likely, zero. The upper limit of the rest mass of ν has been determined from the measurements of the maximum energy of electrons from nuclear β-decay. If the neutrino has a rest mass then it must be less than 60 eV or $<10^{-4}m_e$. The total energy E of a particle with rest mass m_0 moving with velocity v is given by Equation (1.1). If m_0 tends to zero, then v tends to c if the particle energy is finite. If the neutrino mass is zero, it must move with velocity c, just as photons do.

During the years 1962–1963, neutrinos were observed in accelerator experiments in which pions decay (Equation (1.25)) and muons with neutrinos ν_μ and $\tilde{\nu}_\mu$ appear as products. The neutrinos ν_μ and $\tilde{\nu}_\mu$ are different from neutrinos which leave reactors or the Sun. Thus, there are four types of neutrinos:

ν_e – electron neutrino or simply neutrino, as in β^+-decay (2.20), (2.21) or K-capture (Equation (2.22));

$\tilde{\nu}_e$ – electron antineutrino or simply antineutrino, formed during β^--decay (2.19) or μ^--decay (Equation (2.23));

ν_μ – mu-neutrino or neutretto formed in μ^--decay (Equation (2.23))

$\tilde{\nu}_\mu$ – mu-antineutrino or antineutretto formed in μ^+-decay, or π^--decay (Equation (1.25)). Neutrettos and antineutrettos always accompany the formation or decay of μ-mesons.

Neutrinos have no electric and no baryonic charge. They can interact only weakly. Neutrinos have only leptonic charge l and this makes them substantially different from photons. The lepton conservation law (Equation (1.9)) is valid independently for the muon group of leptons (i.e. μ^-, μ^+, ν_μ, $\tilde{\nu}_\mu$) and for the electron group (i.e. e^+, e^-, ν_e, $\tilde{\nu}_e$). If N is the number of leptons, then in all processes the electron leptonic number l_e and independently the muon leptonic number l_μ are conserved:

$$N(e^-)+N(\nu_e)-N(e^+)-N(\tilde{\nu}_e) = \text{constant}; \quad \Delta l_e = 0 \qquad (2.26)$$

$$N(\mu^-)+N(\nu_\mu)-N(\mu^+)-N(\tilde{\nu}_\mu) = \text{constant}; \quad \Delta l_\mu = 0. \qquad (2.27)$$

Neutrinos are helical particles which means that their spin is oriented either parallel to the direction of flight (i.e. to the momentum vector) or the spin is antiparallel to the momentum. The parallel orientation (called also right-handed orientation) of spin and momentum vectors is characteristic for antineutrinos. This is a remarkable property because electrons or protons can move either with right-handed or left-handed orientation.

The interaction of neutrinos with all other particles (including other neutrinos) is extremely weak. Only one of 10^{11} neutrinos is captured on its path from the solar center to the solar surface. The Earth is quite transparent for neutrinos, so that neutrino detectors may be located in deep mines and record the solar neutrinos day and night.

2.3.4. WEAK INTERACTIONS INVOLVING NEUTRINOS

Equation (2.19) can be written in other forms, e.g. (2.20), because the appearance of a particle is equivalent to the disappearance of its antiparticle and vice versa. Then from Equation (2.19)

$$n + \nu_e \rightarrow p + e^- \tag{2.28}$$

and

$$p + \tilde{\nu}_e \rightarrow n + e^+. \tag{2.29}$$

In the second equation the antineutrino must have energy at least 1.8 MeV for the reaction to occur. Comparing Equation (2.28) with Equation (2.19) we see that $\tilde{\nu}_e$ on the right-hand side is equivalent to ν_e on the left-hand side. Another form of Equation (2.19) is

$$n + e^+ \rightarrow p + \tilde{\nu}_e \tag{2.30}$$

and if sufficient energy is available, this reaction may proceed in the reverse direction, which is the process (2.29). Reversal of the arrow means that the exothermic reaction (2.30) becomes an endothermic one (Equation (2.29)). Similarly from Equation (2.28) we get

$$p + e^- \rightarrow n + \nu_e. \tag{2.31}$$

This eaction is called inverse β-decay. It occurs in some atoms when an inner electron is captured by a proton of the nucleus. The proton is changed into a neutron and the atomic number of the nucleus becomes one unit smaller. The process is called orbital electron capture, or K-capture (L-capture), if the shell occupied by the captured electron is known. For example

$$^{37}Ar + e^- \rightarrow {}^{37}Cl + \nu_e \quad (35 \text{ days}; 0.8 \text{ MeV}) \tag{2.32}$$

This reaction is a source of electron neutrinos and in the reverse form it is used for solar neutrino detection.

2.3.5. NEUTRINO PRODUCING REACTIONS

Besides the aforementioned reactions other neutrino producing reactions should be possible according to the weak interaction theory. They are probably important in some evolutionary processes in the universe (e.g. in the Big Bang, in the last phases of thermonuclear evolution of stars etc.). We should mention the following:

Neutrinic bremsstrahlung

$$e^+ + N(Z, A) \rightarrow e^+ + N(Z, A) + \nu_e + \tilde{\nu}_e \tag{2.32a}$$

in which a neutrino-antineutrino pair is formed when a positron is accelerated in the field of the nucleus $N(Z, A)$. Synchrotron radiation of neutrinos in strong magnetic fields

$$e^- \rightarrow e^- + \nu_e + \tilde{\nu}_e \tag{2.33}$$

is analogous to photon synchrotron radiation, but it does not seem to have an important role in the universe.

Electron–electron scattering

$$e^\pm + e^\pm \rightarrow e^\pm + e^\pm + \nu_e + \tilde{\nu}_e. \tag{2.34}$$

Photoneutrino from γ-photons in the field of an electron or a nucleus

$$\gamma + e^- \rightarrow e^- + \nu_e + \tilde{\nu}_e \tag{2.35}$$

and

$$\gamma + N(Z, A) \rightarrow N(Z, A) + \nu_e + \tilde{\nu}_e. \tag{2.36}$$

γ–γ neutrinos

$$\gamma + \gamma \rightarrow \gamma + \nu_e + \tilde{\nu}_e \tag{2.37}$$

or

$$\gamma + \gamma \rightarrow \nu_e + \tilde{\nu}_e. \tag{2.38}$$

Annihilation of an electron–positron pair gives

$$e^+ + e^- \rightarrow \nu_e + \tilde{\nu}_e. \tag{2.39}$$

The pairs are formed from γ photons (Equation (1.17)) with energy higher than 1 MeV. In most cases the electron–positron pair will annihilate into γ-photons (Equation (1.15)), but there is a certain, though very small probability, that reaction (2.39) will occur. Decay of plasmons is given by

$$\Omega \rightarrow \nu_e + \tilde{\nu}_e. \tag{2.40}$$

A plasmon is a quantum of plasma waves, just as photons are quanta of electromagnetic radiation.

Recombination of atoms may be also source of neutrinos

$$e^-(\text{free}) \rightarrow e^-(\text{bound}) + \nu_e + \tilde{\nu}_e. \tag{2.41}$$

Mu-neutrinos (neutrettos) should be, according to the theory, produced by the same processes, e.g.

$$\gamma + \gamma \to \nu_\mu + \tilde{\nu}_\mu \tag{2.38a}$$

$$\gamma + N(Z, A) \to N(Z, A) + \nu_\mu + \tilde{\nu}_\mu \tag{2.36a}$$

etc.

2.3.6. URCA PROCESS

The Urca process consists of two reactions:

$$e^- + N(Z, A) \to N(Z-1, A) + \nu_e \tag{2.42}$$

and

$$N(Z-1, A) \to N(Z, A) + e^- + \tilde{\nu}_e. \tag{2.43}$$

The energy of a very hot plasma is transformed by the Urca process into neutrinos and antineutrinos. For both reactions to proceed, the radioactive isobar $N(Z-1, A)$ must be more massive than $N(Z, A)$ and the energy of the electrons in Equation (2.42) must be very high. By such a process the energy should be drained away from a hot star like money from the tables in Urca Casino, the famous casino in Rio de Janeiro. The Urca process is mentioned here more for historical reasons than for its de facto importance in the energy transport in stars.

2.3.7. NEUTRINO SPECTRUM

The word neutrino spectrum of a source means number of emitted neutrinos as a function of their energy. In general it may consist of continuum and lines (Figure 2.17), just as in a photon spectrum. A neutrino line corresponds to a monochromatic flux, while the continuum represents a broad band of energies. Thus the first reaction in the solar proton–proton chain (see Table 2.3)

$$p + p \to {}^2H + e^+ + \nu_e \tag{2.44}$$

emits neutrinos with a continuum spectrum. The momentum and energy are distributed among three particles, so that the neutrinos may have various direction and energies.

On the other hand, in a process similar to Equation (2.44)

$$p + p + e^- \to {}^2H + \nu_e \tag{2.45}$$

the resulting deuteron and neutrino can move only in opposite directions. The emitted neutrinos are always monoenergetic and represent a spectral line (the kinetic energy of the protons and electron is negligible, 1 keV, in comparison with the energy liberated by the interaction ~ 1 MeV). The line will be slightly broadened by the thermal motion of the interacting particles. The reaction (Equation (2.44)) has a higher probability in the solar interior than the reaction (Equation (2.45)) and the

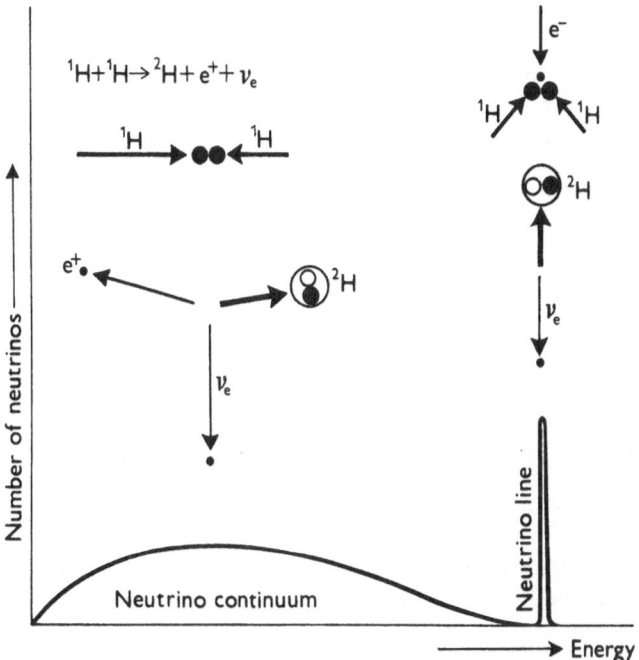

Fig. 2.17. A neutrino spectrum (like a photon spectrum) is a synthesis of continuum and lines. Two neutrino-producing reactions are given as examples. In the first the energy and momentum are divided between three particles (continuum). In the second case the two resulting particles can move only in opposite directions and the neutrino will have the same value of energy (line).

continuum should be more intense therefore in the neutrino spectrum than the line from reaction (2.45).

2.3.8. Cosmic neutrino sources

There are many sources of neutrinos in the universe. From our knowledge of the conditions and processes in cosmic bodies, the emission rate and spectral distribution of neutrinos may be calculated. Whether the predicted neutrino fluxes and neutrino spectrum will correspond to reality is, at present, difficult to verify. Observations of cosmic neutrinos are very difficult. The response of all measuring instruments to neutrino fluxes is extremely poor. This lamentable situation is due to the negligible interaction of neutrinos with the measuring instruments. But, paradoxically, the weak interaction of neutrinos is the cause not only of their misery but also of their grandeur. They bring direct information about processes in the central regions of our Sun and other stars as well. Neutrinos, like photons, transport information and energy. According to our present knowledge, the neutrino luminosity of stars should not be negligible when compared with the photon luminosity. It seems that in advanced stages of thermonuclear evolution of stars, the neutrino luminosity substantially prevails over the photon luminosity.

2.3.9. COSMIC-RAY NEUTRINOS

Interactions of cosmic rays with nuclei of atmospheric atoms produce nuclear fragments, nucleons, pions, muons, electrons, γ-photons and neutrinos. High energy neutrinos and antineutrinos are decay products of kaons, pions (Equation (1.25)) and muons (Figure 2.18):

$$K^+ \rightarrow \mu^+ + \nu_\mu \tag{2.46}$$

$$K^- \rightarrow \mu^- + \tilde{\nu}_\mu \tag{2.47}$$

$$\pi^+ \rightarrow \mu^+ + \nu_\mu \tag{2.48}$$

$$\pi^- \rightarrow \mu^- + \tilde{\nu}_\mu \tag{2.49}$$

$$\mu^- \rightarrow e^- + \tilde{\nu}_e + \nu_\mu \tag{2.50}$$

$$\mu^+ \rightarrow e^+ + \nu_e + \tilde{\nu}_\mu. \tag{2.51}$$

The neutrettos ν_μ and $\tilde{\nu}_\mu$ can be detected by instruments in deep mines. If the energy of neutrettos and antineutrettos is several hundreds of MeV, then they interact with nucleons in the Earth's crust as

$$\nu_\mu + n \rightarrow p + \mu^- \tag{2.52}$$

and

$$\tilde{\nu}_\mu + p \rightarrow n + \mu^+. \tag{2.53}$$

Both μ^- and μ^+ are charged particles and can be easily measured. By analogy Equations (2.28) and (2.29) can be used to detect electronic neutrinos by measuring electrons and positrons.

2.3.10. SOLAR NEUTRINOS

For us, the Sun is the most powerful source of neutrinos. They are liberated in thermonuclear reactions in the small, hot, central part of the Sun. The volume of the neutrino Sun is thus considerably (about 10^6 times) smaller than the volume of the visible Sun. It is comparable with the size of our Earth.

Each elementary proton chain transforms four protons into one He nucleus (Table 2.3). The synthesis proceeds through different steps and at the end two protons must be transformed into neutrons according to reaction (2.20). Thus an elementary process of the He synthesis produces two neutrinos. It yields an energy of 28 MeV $\approx$ 4×10^{-5} erg. The solar luminosity 3.8×10^{33} erg s^{-1} is therefore produced by 3.8×10^{33} erg s$^{-1}/4 \times 10^{-5}$ erg, or 10^{38} elementary transformations per second. The solar neutrino luminosity should then be about $2 \times 10^{38} \nu_e$ s^{-1}. With an average energy of about 0.5 MeV per neutrino, the solar neutrino luminosity represents 10^{38} MeV s^{-1} or 4% of the radiation emitted.

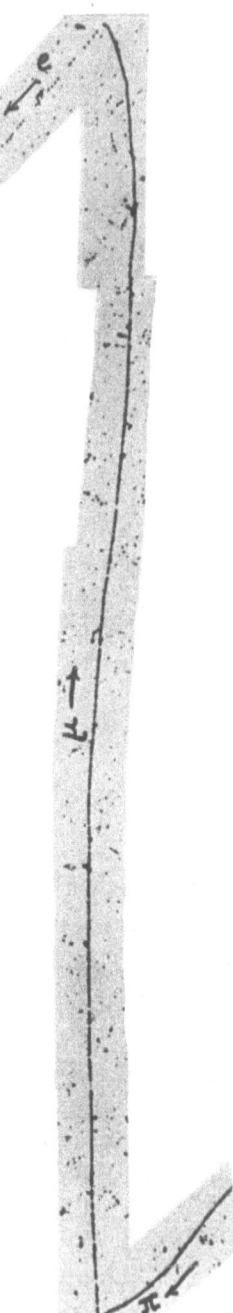

Fig. 2.18.   Decay of a pion (Equation (2.49)) and the following muon decay (Equation (2.50)). Flight of particles is indicated by arrows (C. F. Powell). The neutrinos are not charged and do not therefore ionize atoms of the emulsion.

TABLE 2.3

Neutrino production in the solar interior by the proton–proton chain (according to Bahcall and Sears), and computed capture rate

Reaction	%	Maximum neutrino energy MeV	Neutrino flux at Earth cm^{-2} s^{-1}	Cross section cm^2	Capture rate SNU
$p+p \rightarrow {}^2H+e^+ + \nu_e$	99.75	0.420	6.0×10^{10}	0.0	0.0
or					
$p+p+e^- \rightarrow {}^2H + \nu_e$	0.25	1.44 line	1.5×10^8	1.7×10^{-45}	0.26
${}^2H+p \rightarrow {}^3He + \gamma$					
First branch					
${}^3He + {}^3He \rightarrow {}^4He + 2p$	86				
or					
${}^3He + {}^4He \rightarrow {}^7Be + \gamma$	14				
Second branch					
${}^7Be + e^- \rightarrow {}^7Li + \nu_e$	90	0.861 line	4.5×10^9	2.9×10^{-46}	1.31
	10	0.383 line			
${}^7Li + p \rightarrow {}^4He + {}^4He$					
Third branch					
${}^7Be + p \rightarrow {}^8B + \gamma$					
${}^8B \rightarrow {}^8Be + e^+ + \nu_e$	0.02	14.06	5.4×10^6	1.35×10^{-42}	7.28
${}^8Be \rightarrow {}^4He + {}^4He$					

The losses by absorption are quite negligible (about one ν_e of $10^{11}\nu_e$ is absorbed when passing from the center to the surface of the Sun). On the Earth the ν_e flux is then $2 \times 10^{38}\nu_e \, s^{-1}/4\pi(AU)^2 = 6 \times 10^{10}\nu_e \, cm^{-2} s^{-1}$. The greatest number of these neutrinos is produced by the p–p reactions (2.44) (Figure 2.19) which yield a continuous spectrum with maximum energy 0.42 MeV. This is unsufficient to be recorded by present neutrino experiments, the energy threshold of which is 0.82 MeV.

For neutrino detection one uses, in principle, the process (2.28), which for a nucleus represents an inverse electron capture:

$$N(Z, A) + \nu_e \rightarrow N(Z+1) + e^-. \tag{2.54}$$

The daughter nucleus $N(Z+1, A)$ is radioactive and can be detected by modern methods. By measuring the rate of the production of $N(Z+1, A)$ nuclei, one determines the rate of ν_e-capture, which in turn depends on the capture cross section and on the flux of ν_e.

The choice of the detecting element $N(Z, A)$ in reaction (2.54) depends on the energy threshold of the reaction, on the half-life of the radioactive element formed $N(Z+1, A)$, on the availability and price of the detecting material etc. Let us

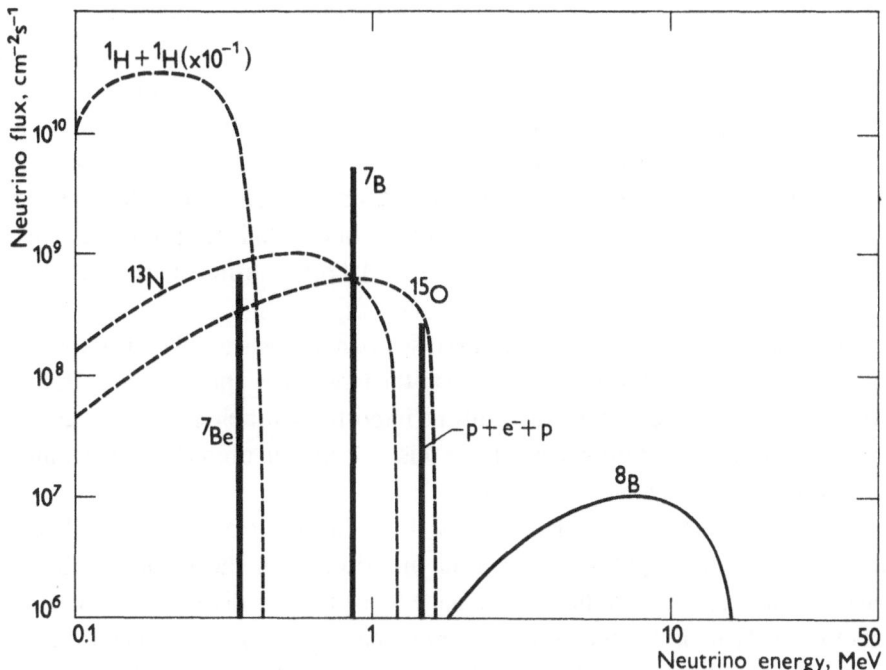

Fig. 2.19.   Computed neutrino spectrum from thermonuclear reactions in the solar interior. Fluxes for continua are in number $cm^{-2} s^{-1} MeV^{-1}$; for lines number $cm^{-2} s^{-1}$ (according to Bahcall).

mention some possibilities:

$$^{87}Rb(\nu_e, e^-)\,^{87}Sr, 0.12\,MeV, 8\,h;$$

$$^{55}Mn \rightarrow\,^{55}Fe, 0.23\,MeV, 2.7\,yr;$$

$$^{71}Ga \rightarrow\,^{71}Ge, 0.23\,MeV, 11\,d;$$

$$^{51}V \rightarrow\,^{51}Cr, 0.75\,MeV, 28\,d;$$

$$^{37}Cl \rightarrow\,^{37}Ar, 0.82\,MeV, 35\,d;$$

$$^{7}Li \rightarrow\,^{7}Be, 0.86\,MeV, 53\,d.$$

For practical reasons the stable isotope ^{37}Cl has been chosen for the present experiments, so that reaction (2.54) in this particular case is

$$^{37}Cl + \nu_e \rightarrow\,^{37}Ar + e^-. \tag{2.55}$$

For detection of the radioactive Ar a reverse process is used, i.e., orbital electron capture reaction (2.32). The experiment is placed in a deep mine in South Dakota, as protection against the direct influence of cosmic rays. The detecting fluid is drycleaning perchlorethylene C_2Cl_4, 3.9×10^5 l of which are in a large tank. But even in such a large amount of the Cl atoms, the capture of a neutrino is a very rare event.

It is usual to specify the rate of the neutrino capture reaction (2.55) in terms of so-called solar neutrino units (SNU or 'snew') which are such that 1 SNU corresponds to 10^{-36} captures per target Cl atom per second. For the amount of C_2Cl_4 used, this corresponds to the production of one ^{37}Ar atom every three days.

Theoretical rates depend on the assumptions made about the conditions in the solar interior and on the laboratory data on reaction rates. In any case the theoretical results are several times larger than the measured rates. While the predicted rates are 6 to 9 SNU, the rates measured with the South Dakota instrument are about 1 SNU or less.

Before assessing the importance of this discrepancy, one should realize that much of the predicted rate(see Table 2.3) is due to the most energetic neutrinos. But these are produced by a process with practically no importance for He synthesis and energy production. If no contribution came from this process, the predicted rate would not be inconsistent with the measurements.

If the discrepancy really exists, it may lie within the experiment itself, in the nuclear physics data applied to the solar interior, or in the model of the solar structure. Many suggestions have been made to explain the discrepancy. There is no lack of ideas. What is needed more is improvement in the observational techniques and, eventually, the use of other target atoms from the list above.

2.3.11. STELLAR NEUTRINOS

Both the photon and neutrino radiations coming to our Earth from stars are much smaller than those from the Sun. In the main sequence stars, which represent the major part of the stellar population in the solar neighborhood, the energy is produced by the transformation of H into He, as in our Sun. A steady flux of neutrinos should leave the stars and flow into cosmic space. A rough estimate of neutrinos produced by hydrogen burning in the universe can be made:

$$\varrho(\nu_e) = L_\nu n_s n_g \tau_u V^{-1} \sim 1.5 \times 10^{-3} \text{ eV cm}^{-3} \tag{2.56}$$

where L_ν is the neutrino luminosity of our Sun $\sim 10^{38}$ MeV s^{-1}, n_s is the number of stars in a galaxy (about 10^{11}), n_g is the number of galaxies in the universe (about 3×10^9), τ_u is the age of the universe (about 4×10^{17} s), V is the volume of the universe ($\sim 10^{85}$ cm^3) (all the aforementioned values are very rough estimates). The mean density of observable matter amounts to 3×10^{-31} g cm^{-3} or in terms of energy density 1.7×10^2 eV cm^{-3}. One can conclude that the energy density of neutrinos produced by nucleosynthesis in stars is about five orders of magnitude smaller than the mean density of observable matter.

The neutrino emission during later stages of stellar evolution, when central temperatures reach values $>10^9$ K, will be governed by reactions (2.32a) to (2.41), and in particular by annihilation of the electron–positron pair reaction (2.39). In the last stages of thermonuclear evolution of the massive stars, neutrino luminosity should prevail over their photon luminosity. In the last phase of stellar life, when the material in the core can yield no further thermonuclear energy, the gravitational implosion of the core starts. Energy losses from the core via the aforementioned neutrino-producing reactions and also by neutronization of protons (Equation (2.31)) at densities $>10^7 \, \mathrm{g\,cm^{-3}}$ (see Section 3.7.3) give rise to a huge neutrino luminosity of the collapsing star – for example of a supernova. Due to the very high luminosities, in particular the neutrino luminosity, a supernova represents a good prospect for observational verification of our present ideas on the stellar agony.

The processes just preceding the core collapse and the stellar envelope explosion (see Section 5.4.3e) left their traces in the chemical abundances of the most stable elements, viz. Fe, V, Cr, Mn, Co, Ni. Their relative amounts have remained unchanged since then. The observed abundances in our solar system correspond well to the theory, thus supporting the idea of huge neutrino luminosities from supernovae.

2.3.12. INTERSTELLAR NEUTRINOS

Cosmic ray particles collide with nuclei of interstellar atoms and their strong interactions produce π-mesons and K-mesons. Both types of mesons decay, resulting in neutrinos with very high energies. Their spectrum, ranging to 10^{17} eV, may be deduced from the spectrum of the primary cosmic rays and from the properties of interstellar matter. In the entire Galaxy, the production of interstellar neutrinos (with energies >10 MeV) represents about 10^{44} neutrinos per second. This is a much smaller contribution (about five orders of magnitude lower) than the neutrino production by thermonuclear processes in stars.

2.3.13. RELIC NEUTRINOS

In addition to the ever increasing neutrino population produced in stars and in the interstellar medium, there should be a large quantity of relic neutrinos (a neutrino sea). If the hypothesis of the Big Bang is correct, then a huge amount of neutrinos and antineutrinos should have been produced in the very early stages of the universe, i.e. in its lepton era. The leptons (positive and negative muons, electrons, positrons, neutrinos, antineutrinos, neutrettos and antineutrettos) were exceptionally abundant in the era when the temperature of the expanding universe was decreasing from 10^{12} to 10^{10} K. The lepton era started when the universe was about 10^{-2} s old and lasted for about 1 min (see Sections 5.2.b). It ended by complete detachment of neutrinos from matter.

The remnant and intact neutrino sea today would have a temperature of about 2 K. Neutrinos with such a low energy cannot be detected by the methods of contemporary physics, even if they exist in great quantities. Thus, the relic neutrinos remain hypothetical particles. The idea, however, received support with the discovery of relic microwave photons which are probably of Big Bang origin. The relic neutrinos would be the oldest neutrinos in the universe.

Summary

All elementary particles are endowed with the weak interaction. Its range, 10^{-15} cm, is much shorter than is that of the strong interaction (10^{-13} cm). Its field particle, the W-meson, should be heavier than the K-meson and may be even heavier than the proton. In weak interactions strangeness and parity are not conserved, so that the weak decay of some particles, e.g. strange particles, K-mesons, π-mesons and μ-mesons, are prohibited and are therefore extremely slow processes (thirteen orders of magnitude slower than strong decays).

Neutrino producing reactions are another type of weak interaction. There are four different types of neutrinos: electron neutrino and electron antineutrino (ν_e, $\tilde{\nu}_e$) which are involved in reactions with electrons or positrons; muon neutrino and muon antineutrino (called also neutretto and antineutretto – ν_μ, $\tilde{\nu}_\mu$) which occur in elementary particle reactions involving muons.

There are different neutrino sources in the universe, such as beta radioactivity of bodies in the solar system, cosmic ray neutrinos from the decay of pions and muons in our atmosphere, thermonuclear reactions in the Sun and other stars, neutrino-antineutrino pair formation during the last evolution phase of massive star evolution and interstellar neutrinos produced in the decay of kaons, pions and muons.

None of the known elementary particles has such a weak interaction with other particles as has the neutrino. The weak interaction of the neutrino is its glory and its misery. Once born it moves through the universe for ever. It is our only hope for observing the central parts of stars directly. After neutrino detectors have been sufficiently developed, it will be possible to decide whether antimatter exists in large quantities in the universe or not: while stars of normal matter (koinomatter) are sources of electron neutrinos, the stars of antimatter should produce electron antineutrinos by their thermonuclear reactions. This problem cannot be solved using photon detectors, because photons and antiphotons are identical particles and hence indistinguishable.

2.4. Gravitation

The fourth, and by far the weakest, force binding elementary particles together is gravitation. In the laboratories where elementary particles and their interactions are studied, gravitation has in the past been nearly neglected. Its importance for the structure of atomic nuclei and atoms is negligible when compared with the other three interactions. The Coulomb force e^2/r^2 is similar to Newton's gravitation $Gm_p m_e/r^2$ and there is also a resemblance between the electronic envelope of an atom and the planetary system of our Sun. However, the gravitational force binding an electron to a proton is much weaker than their electric attraction, viz $Gm_p m_e/e^2 \sim 10^{-40}$. That is the reason why gravitation is neglected in atomic physics. On the other hand, gravitation is a universal interaction, acting between all types of elementary particles. Even photons, the quanta of electromagnetic fields, experience it.

2.4.1. LARGE-SCALE STRUCTURAL FORCE

If gravitational forces are negligible in the microworld when compared with the other interactions, they are of paramount importance on cosmic scale. Strong interactions have a very small range, 10^{-13} cm, and weak interactions reach even to smaller distances, $<10^{-15}$ cm. Electric forces could act over large distances but any matter in larger quantities has a strong tendency to be neutral. Electrons shield the electromagnetic action of protons. On the other hand, amounts of gravitational attraction increases with increasing amounts of matter and there is no shielding. In a cosmic body, the large number of all elementary particles present participate in creating the total gravitational field of the body. Gravitational attraction between two bodies (e.g. the Earth and the Moon) is the sum of the gravitational interactions between any particle of one body and all the constituent particles of the other body. There is no screening, as is the case for electric forces.

Gravitation is a universal force in the universe. Let us recall a few examples. Our body would fly away from the Earth into cosmic space if it were not for the gravitational attraction of each of its elementary particles and all the particles constituting the Earth. The flow of rivers, the winds, tides, fall of objects and their weight, the motion of the Moon and of man-made satellites are all consequences of the Earth's gravitation. Convection in terrestrial atmosphere and hydrosphere, in the subphotospheric layer of the Sun and of other stars too, is controlled by gravitation. The motion of satellites around their planets; the revolution of planets and comets around the Sun; the trajectories of space probes; the motions of stars in binaries, in multiple stars and in stellar clusters; the revolution of the solar system, nebulae, stars, interstellar matter and stellar clusters around the galactic center, the motion of our Galaxy in the Local cluster of galaxies and the deceleration of the expansion of the universe, all these are a few examples of gravitational interaction on different scales.

2.4.2. SELF-GRAVITATION

Gravitation governs not only motions of celestial bodies and of their systems, but it also determines their shape.

The self-gravitation of small bodies, like meteorites (meteoroids), small planetoids and little satellites, is insufficient to overcome the strength of solids, caused by electromagnetic interactions. Solids yield or fracture under a tension or a load of $\sim 10^9$ dyn cm^{-2} or $\sim 10^3$ atmospheres. A column $L\varrho$ of material (L is length, ϱ is density and the cross section of the column is 1 cm^2) would produce a load of $L\varrho g_E$ on the Earth ($g_E \approx 10^3$ cm s^{-2}). A column of height $L \sim 10$ km would produce the critical load. Higher structures would yield under their own weight.

On the surface of another planet with radius R and mean density ϱ the gravity acceleration is

$$g = GMR^{-2} = \frac{G4\pi R^3 \varrho}{3R^2} = \frac{4\pi}{3} GR\varrho. \tag{2.57}$$

Thus

$$\frac{g}{g_E} = \frac{R\varrho}{R_E\varrho_E}.$$

For equal weights

$$L\varrho g = 10 \text{ km}\varrho_E g_E. \tag{2.58}$$

The material of a solid body with the radius R will yield under self-gravitation, if $R \sim L$ or $R \approx 6 \times 10^2 \sqrt{\varrho_E}/\varrho$ km. For a mean density of $\varrho = 3.4$ g cm^{-3} (density of the Moon) the critical radius would be roughly 400 km, which is the size of the largest asteroids. We may conclude that the largest asteroids and largest satellites will be shaped by gravitation.

Unless influenced by a very fast rotation or by the presence of another massive body in the immediate neighborhood, stars and planets have a spherical shape due to self-gravitation. It is the shape with minimum potential energy.

Stars and stellar systems are distorted by the gravitation of nearby bodies. Thus, the binary β Lyrae has its components elongated by mutual gravitational attraction into the shape of ellipsoids. In close binaries the surface material of the expanding component is driven by gravitation to its companion and large amounts of stellar material may be transferred in this way.

Galactic star clusters (called also open clusters) consist of a small number of stars (10^2 to 10^3); so that the self-gravitation of the cluster is small and its stars are torn away by the gravitational forces of the Galaxy. A surprisingly large number of galaxies appear connected by luminous intergalactic bridges, which may be remainders of enormous tides. Morphological anomalies of some peculiar galaxies are apparently after-effects of gravitational forces exerted during close passage of galaxies. Also spiral arms have sometimes been explained as tidal ejections and counter-ejections, later deformed by differential rotation. The idea of tidal deformation of galaxies has been verified by computer-simulated close passages of galaxies.

Depending upon the number of particles held together by self-gravitation, a cosmic body is either a self-luminous star with 10^{55} to 10^{59} elementary particles, or a planet with less than 10^{54} particles.

Gravitation is of basic importance for the structure and shape of planets, stars and of their systems. It is also of basic importance in their formation, evolution and death. Stars were formed by gravitational contraction of interstellar matter (see Section 5.4). The gravitational energy of the contracting protostar is partly radiated away and partly used for heating its central region (virial theorem). For the least massive stars (featherweight stars), gravitation is the principal source of luminosity. If such stars reach the stage of thermonuclear burning at all, then they burn only the minor fuels, such as D and Li.

In the more massive stars ($M > 0.08 \, M_\odot$) the central temperatures are sufficient for H burning. In later phases of stellar evolution, after H has been converted into

He, the stellar core is again compressed by gravitation and its temperature increases to about 10^8 K. At this temperature, He conversion into C sets in. The star is now a red giant. During further evolution gravitation sets in for short periods to replace thermonuclear reactions and to raise the temperature of central regions so that further reactions may be ignited. After exhaustion of all thermonuclear fuel in the stellar core, self-gravitation causes its collapse.

The aforementioned gravitational interactions in the universe lead to a mass defect. Mass is a non-additive property. If two stars with masses M_1 and M_2 merge together and the total baryon number is unchanged, the mass of the resulting star M_3 is such that

$$M_1 + M_2 > M_3. \tag{2.59}$$

The decrease

$$M_1 + M_2 - M_3 = \Delta M \tag{2.60}$$

is called the gravitational mass defect. It would be radiated away by the new star in the form of gravitational waves, photons and neutrinos. By analogy, the mass of a star is less than the mass of the protostellar cloud from which it condensed, even if no baryons were to escape. The energy

$$E = (\Delta M)c^2 \tag{2.60a}$$

is radiated away into interstellar space. The gravitational mass defect is a measure of stability: the larger the mass defect the larger is the *binding energy* of all particles in the star or of the stars in a stellar system.

2.4.3. NEWTONIAN THEORY OF GRAVITATION

In his *Principia* Newton described gravitation as a cause that acts on the Sun and planets according to the quantity of matter which they contain and propagates in all directions to immense distances, decreasing always as the inverse square of the distance. A vague idea about the decrease of gravitation with distance had already been pronounced in the ninth century (Johannus Scotus Erigena 815–877) and the inverse-square dependence had been suggested in 1640 (Imael Bulliardus). But Newton was the first who deduced the inverse-square law from observations, viz. from Kepler's laws, which in turn were deduced from Tycho Brahe's observations.

The Newtonian law of gravitation states that every particle in the universe attracts every other particle with a force acting along the line joining the particles. Their masses m_1 and m_2 at positions $\mathbf{r}_1$ and $\mathbf{r}_2$ attract each other with the forces

$$\mathbf{F}_1 = -\frac{Gm_1 m_2}{|\mathbf{r}_1 - \mathbf{r}_2|^3}(\mathbf{r}_1 - \mathbf{r}_2)$$

and

$$\mathbf{F}_2 = -\frac{Gm_1 m_2}{|\mathbf{r}_2 - \mathbf{r}_1|^3}(\mathbf{r}_2 - \mathbf{r}_1) \tag{2.61}$$

where G is the gravitational constant 6.7×10^{-8} dyn cm^2 g^{-2}. If more particles are involved the gravitational force on the nth particle is described as

$$\mathbf{F}_n = m_n \frac{d^2 \mathbf{r}_n}{dt^2} = G \sum \frac{m_m m_n (\mathbf{r}_m - \mathbf{r}_n)}{|(\mathbf{r}_m - \mathbf{r}_n)|^3}. \tag{2.62}$$

The law of gravitation has explained many phenomena, e.g. the free fall of bodies, revolution of the Moon, revolution of planets (i.e. Kepler laws) etc. The masses of the Earth, Sun and planets are determined from orbits of satellites around their planets, or orbits of planets around the Sun; eventually also from orbits of comets or space vehicles which passed sufficiently close to the planet or other body of the planetary system whose mass is to be determined.

The masses of binary stars may be deduced only from their gravitational interactions. The mass of the Galaxy may be deduced from galactic orbits of stars. The Galaxy does not rotate as a solid body; its rotation is differential. That means that the stars revolve around the galactic center with angular velocities dependent on their distance from the center. The distribution of mass with the distance in the Galaxy can thus be determined.

Newton's law of gravitation met with many successes. There were irregularities discovered in the orbit of Uranus, but they led to the prediction and discovery of Neptune – which was the outstanding verification of Newton's theory. Notwithstanding, one problem persisted unsolved, viz. the observed precession of the Mercury perihelion was 43″ per century faster than expected from gravitational perturbations by other planets, when the Newtonian theory was used. Just this discrepancy in the Newtonian theory was the first confirmation of the Einstein theory of gravitation (general theory of relativity).

2.4.4. GEOMETRICAL INTERPRETATION OF GRAVITATION

According to the Newtonian law, gravitational effects should be instantaneous. Thus the change of position $\mathbf{r}_2$ in Equation (2.61) should be felt immediately by the particle m_1, whatever the distance between m_1 and m_2 may be, since neither velocity nor time are involved. One would therefore be able to send information with infinite speed by dislocating some mass. On the other hand, the theory of relativity gives arguments that no mass nor energy, and hence no information, can move faster than the velocity of light (compare e.g. formula (1.1)). The finite velocity propagation of gravitation is taken into account in the Einstein law of gravitation (Einstein field equations).

The Einstein law does not only mean a substitution of Newton's law by a better one. The notions of space and time, and thus the whole framework of the newtonian physics were changed simultaneously (Table 2.4). In general relativity the concepts of curvature of space-time and gravitational field are equivalent (geometrization of physics). Mass and energy of any particle and of any field are also sources of gravitation i.e. of curvature of space-time.

TABLE 2.4

Differences between the Newtonian theory of gravitation and the general relativity

	General relativity	Newton theory
described by	ten metric potentials $g_{\mu\nu}$ determining geometrical properties of space-time	one potential $-GM/r$
space-time	curved in presence of gravitation	Euclidean
source	mass-energy of any particle and of any field	mass of particles
theory	non-linear: gravitation field of more bodies is not a simple sum of effects of individual bodies	linear theory: effect of several bodies is a simple summation
propagation	with velocity c – gravitational waves	immediately at any distance, infinite velocity

At present there exist different theories concerning gravitation: General relativity of Einstein, scalar-tensor theory of Brans-Dicke, vector-tensor theory, tensor-tensor theory and others. A great majority of physicists and astronomers prefer the general theory of relativity for its beauty, simplicity, and its agreement with observations and with experiments.

In the theory of relativity the time and space coordinates are not considered separately, because the length and time intervals depend on the relative motion of the observer. Space and time coordinates form a four dimensional space-time continuum, one point of which is called an event. In the Newtonian physics the space and time are a kind of absolute stage in the framework of which all physical processes take place. On the contrary, in general relativity the physical events are closely tied to space-time, the properties of which depend upon the distribution of matter and energy. Therefore the gravitation in general relativity is explained by geometrical properties of the space-time, i.e. as curvature or 'warping' of space and time by matter and energy.

The information about the space-time geometry is given by the metric. It expresses the space-time 'distance' ds between one event and its neighbor. An event in space-time is analogous to a point in a three-dimensional space. It has no dimensions and no duration. Its three coordinates fix its position in space relative to some frame of reference. The fourth coordinate gives the instant of occurrence of the event in the same frame of reference. The space-time distance or separation ds- called more appropriately the interval between two events x_1, x_2, x_3, x_4 and $(x_1 + dx_1), (x_2 + dx_2),$ $(x_3 + dx_3), (x_4 + dx_4)$ is a combination of space and time differences dx_1, dx_2, dx_3, dx_4:

$$ds^2 = \sum_{i,\,k=1}^{4} g_{ik}\, dx^i\, dx^k \tag{2.63}$$

In a flat space-time, the interval is given by the Pythagorean theorem

$$ds^2 = dx_1^2 + dx_2^2 + dx_3^2 + dx_4^2 \tag{2.64}$$

which is a particular case of Equation (2.63).

The g_{ik} in Equation (2.63) are called coefficients (or components) of metric and they are functions of space-time coordinates. Their mathematical form depends not only on the space-time geometry but also on the choice of the frame of reference. If another frame is used for the same space-time, the g_{ik} are different. They transform as a tensor. While components of the metric tensor g_{ik} depend on the coordinate system used, the interval ds remains unchanged. In other words, the interval between two neighboring events is invariant under coordinate transformation. The relation (2.63) is called metric of the space. A metric is the generalization of the Pythagorean theorem including time, with variable coefficients and for small distances. It is important that the system of 16 coefficients g_{ik} represents a tensor, because rules for transforming tensors from one frame of reference to another are well known. Since $dx_i\, dx_k = dx_k\, dx_i$, the metric tensor is symmetric and there are only 10 independent components g_{ik}.

The manner in which the curvature of space-time is produced by the material content, i.e. the relation between the coefficients g_{ik} and the mass and energy is given by Einstein's law of gravitation (Einstein's field equations). It has a form of ten equations relating the metric coefficients g_{ik} and their derivatives (on the left-hand side) with the distribution of mass and energy (on the right-hand side). How a particular distribution of matter and energy will shape the space-time can be found by solving the Einstein's equations. Solving the equations is, however, a difficult mathematical task. The distribution of matter in the universe is very complex and only simplified models of the distribution are tractable. Some symmetry in the distribution of matter is generally assumed (e.g. a spherical star in an empty space or uniform distribution of matter in the universe) and the mathematics involved becomes considerably simpler.

In the planetary system with weak gravitational forces both laws of gravitation are in good agreement. There are, nevertheless, three phenomena which the Newton theory cannot explain and general relativity can: the deflection of star light in the solar gravitational field the gravitational redshift and the perihelion advance of planets.

(a) Deflection of Radiation

The light of a star seen at apparent distance R from the center of the solar disk is predicted by the general relativity theory to be deflected by $1.75'' \, R^{-1}$; apparent solar radius $959.6''$ is used as unit for R. The measurements during total solar eclipses give values in the range 1.75 to 2.2 arc seconds. Recently, long-based radio interferometers (two or more radio telescopes several thousand kilometers apart) were used to measure the deflection of radio waves passing close to the Sun. Angles as small as $0.0003''$ may be resolved. Two quasars (3C 273 and 3C 279) are close to the Sun on October 8 each year (Figure 2.20). By long-based interferometry the radio astronomers may determine on that day how much the radio rays from both quasars are bent with respect to each other and thus determine whether the general theory of relativity is the correct metric theory of gravitation or not. The values measured to date (close to 1.75″) agree with the Einstein theory, but the values predicted by some other theories still lie within the margin of errors.

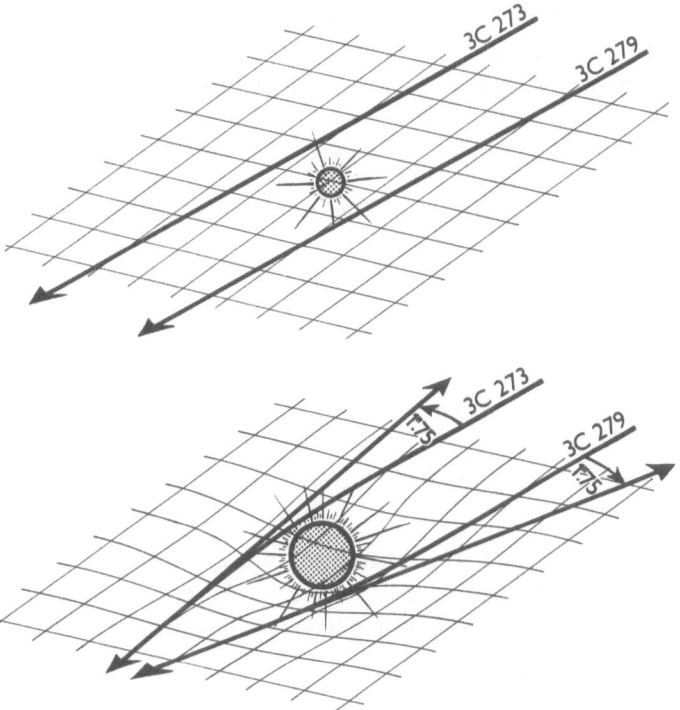

Fig. 2.20.   Newtonian space around massive bodies remains euclidean, so that the rays should not change their direction. Observed deflection of rays however indicates that the space is curved. The warping represented in the figure is two-dimensional only.

(b) *Gravitational Redshift*

A solar photon incident on our Earth has traversed a potential difference $GM_\odot/R_\odot$ and its energy should be decreased by

$$\Delta(h\nu) = -\frac{GM_\odot h\nu}{R_\odot c^2}. \tag{2.65}$$

According to Equation (1.3) $h\nu/c^2$ is the equivalent mass of the photon. The decrease of the frequency $\Delta\nu$ means a redshift in the spectrum $\Delta\lambda$. This is equivalent to a Doppler velocity $c(-\Delta\nu/\nu) = c(\Delta\lambda/\lambda)$ (as $\Delta\nu = -c\Delta\lambda/\lambda^2$), numerically equal to

0.64 km s^{-1}. Though qualitatively confirmed, the quantitative measurements for the Sun and for white dwarfs are difficult.

However, the gravitational redshift has been measured on our Earth using the Mössbauer effect on very sharp γ-lines. The gravitational redshift of the photons falling in the tower of the height of ~ 10 m was measured with a very high accuracy, $\Delta \nu / \nu = 10^{-16}$.

(c) *Perihelion Advance of the Planets*

The third classic test of the Einstein law of gravitation is the advance of the perihelion of planets. The values predicted by the theory and the values deduced from observations are in a very good agreement, as may be seen from Table 2.5.

TABLE 2.5

Theoretical and observed values of perihelion advance, expressed in arcseconds per century

	Relativity	Observation
Mercury	43.03″	43.11″
Venus	8.6	8.4
Earth	3.8	5.0
Icarus	10.3	9.8

All three tests are a remarkable verification of the general relativity within our planetary system. The expansion of the universe is the fourth and more dramatic test on a large scale. Straight from the Einstein law, Friedmann deduced (in 1922) that a closed universe of uniform density will inevitably expand, reach a maximum dimension and contract again, ending with a final gravitational collapse. It is worth noting that Hubble confirmed the expansion of the universe by his observations of distant galaxies five years later. The importance of general relativity for the structure of the universe at large scale is discussed in Section 4.9.2.

An entire series of new experiments has been planned to test the metric theories of gravitation. Many of them must be carried out by means of spacecraft.

2.4.5. STRONG GRAVITATIONAL FIELDS

Strong gravitational forces act inside and close to very dense cosmic bodies. The curvature of the space-time there and deviations from the Newtonian laws are considerable. Such a situation occurs at the final evolutionary stage of middleweight and heavyweight stars, when they collapse into neutron stars and black holes.

A black hole is formed when gravitation on the surface is such that no photons or any other particles can escape from it. This possibility was known in Newtonian physics about two hundred years ago. Laplace put forward the following argument.

For a particle to escape from the surface of a spherical body to infinity, its kinetic energy must be higher than its potential energy, viz.

$$\tfrac{1}{2}v^2 > \frac{GM}{R},\tag{2.66}$$

where M and R are the mass and radius of the body. The particle cannot escape if

$$\tfrac{1}{2}v^2 < \frac{GM}{R},$$

because its energy is not sufficient to bring it out of the gravitational potential of the body. If the radius of the body is $R < 2GM/c^2$ a particle cannot escape even if it moved with the velocity of light. The same value follows from the general relativity for the so-called gravitational radius (or Schwarzschild radius) R_g, viz.

$$R_g = \frac{2GM}{c^2}\tag{2.67}$$

of a spherical body with mass M. From the Schwarzschild surface with the radius R_g, the radiation cannot escape to infinity, so that the surface represents an event horizon. In the interior of the Schwarzschild surface the density is so high that the known physical laws may be not valid there.

The limit (Equation (2.67)) follows from the metric ds deduced by Karl Schwarzschild (1916) for a spherical mass surrounded by a vacuum. A more general solution of the Einstein equations was found by Kerr (1963). The Kerr metric is also for vacuum but around rotating stars. The Kerr solution has two possibilities, according to the parameters of the star: a singularity that may be visible from outside (no event horizon, a naked singularity) or a solution with event horizon. In the latter case, an additional surface exists (stationary limit) which is external to the event horizon and touches it at the poles. The observer at stationary limit must move with the velocity of light to remain there and the light emitted by him will arrive infinitely redshifted to an observer at infinity.

The space between the event horizon and the stationary limit is called the ergosphere. The coordinate t (time-like outside the stationary limit) becomes space-like in the ergosphere. This transformation leads to the possibility of extracting energy from a rotating black hole (as proposed by Penrose in 1972). A piece of matter (e.g. a star) which encounters a black hole is split into two parts in the ergosphere and due to the transformation, one part can assume negative energy and falls into the hole. The other part increases its original energy (due to energy conservation) and escapes to infinity with an excess of energy. The extraction of the rotational energy from a Kerr black hole might explain explosive processes in galactic nuclei. A supermassive collapsed body (black hole) with mass of $\sim 10^8\,M_\odot$ is supposed to rotate in an active galactic nucleus. But this is a pure hypothesis, and other explanations of active galactic nuclei exist (see Sections 4.7c,d).

2.4.6. GRAVITATIONAL WAVES

There is another important phenomenon which follows directly from the Einstein field equations, i.e., gravitational radiation. Any nonspherical, dynamically changing system must emit gravitational waves.

Gravitation (in particular, weak gravitational fields) is in some respects similar to electric forces, and thus gravitational waves also have properties similar to electromagnetic waves. Both forces are perpendicular to their propagation and both carry energy at the speed of light; their amplitude falls off as the inverse of the distance while their flux decreases as the inverse square of the distance. An electromagnetic wave accelerates charged particles in the plane perpendicular to its propagation. A gravitational wave passing through a cloud of particles also accelerates them perpendicularly to the wave propagation. But there is also a relative acceleration of particles (Figure 2.21).

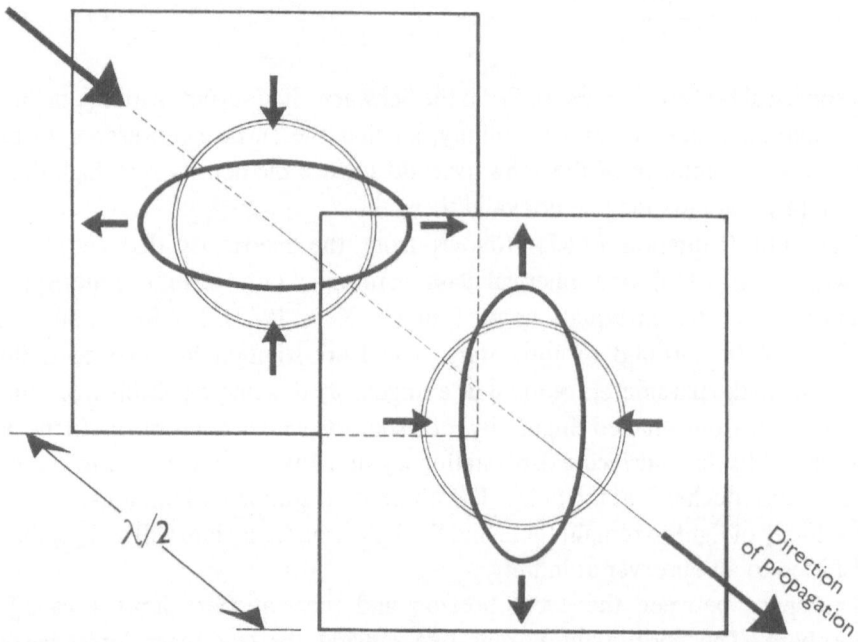

Fig. 2.21.   Forces of a gravitation wave in the plane perpendicular to the direction of propagation. They are marked by short heavy arrows. Circular distribution of mass particles (double circle) is deformed by the incident gravitational wave to a shape of the heavy line. The change after half a period ($\lambda/2$) is seen in the foreground plane.

The gravitational radiation can be emitted by a system with a changing quadrupole moment, such as rotating stars without axial symmetry, binary stars, non-radial pulsations of stars, explosion and collapse of stars, if the process is not axially symmetric. In all the cases the distribution of mass can be approximated by a quadrupole.

The Earth emits about 10 W in the form of gravitational waves due to its nonaxially distribution of masses. Two neutron stars (with mass $\sim\frac{1}{2}M_\odot$ and $R \approx$ 10 km) revolving around each other at a distance of 100 km should emit about 10^{42} W, which represents a substantial energy loss for the system. Such a system would lose its energy in a very short time and collapse. Any binary star with a very short period $(P < 1\,h)$ has little chance to survive for long due to the strong gravitational wave losses. Most existing binaries are much wider and their gravitational wave radiation correspondingly smaller.

The flux of gravitational wave energy coming to the Earth depends also on the distance of the emitting source. For example the flux from ι Bootis is $10^{-10}\,\mathrm{erg\,cm}^{-2}\,\mathrm{s}^{-1}$, with frequency $\nu = 7.5$ per day (period 3.2 h). Other binaries are fainter and their fluxes are smaller, 10^{-11} to $10^{-12}\,\mathrm{erg\,cm}^{-2}\,\mathrm{s}^{-1}$. The total flux from all the binaries in our Galaxy with periods of a few hours is about $10^{-7}\,\mathrm{erg\,cm}^{-2}\,\mathrm{s}^{-1}$. The detectors used today are sensitive to 10^{-1} to $10^{-2}\,\mathrm{erg\,cm}^{-2}\,\mathrm{s}^{-1}$. The future detectors may have sensitivity 10^{-7} to $10^{-8}\,\mathrm{erg\,cm}^{-2}\,\mathrm{s}^{-1}$ and there exists a possibility of detection of gravitational waves from galactic binary stars.

A suggestion has been made that gravitational waves might be produced in a huge black hole (with mass of millions of $M_\odot$) in central parts of our Galaxy. Whenever a star of solar mass would be swallowed by the black hole, an energy of 10^{45} erg would be emitted as gravitational waves, within minutes. It would provide a flux of about $10^{-3}\,\mathrm{erg\,cm}^{-2}\,\mathrm{s}^{-1}$ at the Earth.

Gravitational waves may be emitted by fast rotating pulsars exhibiting some departures from axial symmetry. Estimates give the highest flux for the Crab pulsar, i.e. $\sim 10^{-13}\,\mathrm{erg\,cm}^{-2}\,\mathrm{s}^{-1}$ at a highly monochromatic frequency 60 Hz (one rotation radiates two waves). Other pulsars should be a few orders of magnitude fainter.

The supernova event leading to a pulsar should also emit gravitational waves, because it probably lacks also axial symmetry. The resulting neutron star has a large gravitational binding energy (a few percent of Mc^2) and a considerable fraction of the energy is probably emitted as gravitational waves during the collapse. The resulting gravitational radiation bursts, with frequencies of kHz and of duration less than one second, should be of the order of magnitude of watts to kilowatts per cm^2 at our Earth depending on the supernova itself and its distance from us. The supernova event is a rare event (roughly one supernova in a galaxy per century). One could observe gravitational waves of supernovae of all nearby galaxies. The radiation fluxes should be from ergs to thousands of ergs, depending on the distance of the galaxy.

Other sources of gravitational radiation have been considered, such as primordial gravitational radiation (analogy to the 3 K isotropic background), condensation of galaxies out of the expanding primordial gas, black holes in globular clusters, dense clusters of black holes (with close encounters between holes radiating bursts of gravitational waves), conversion of electromagnetic waves into gravitational waves in static electric and magnetic fields. Detailed discussion of all the emission processes may be found in numerous papers.

Different systems for detection of gravitational waves have been used or proposed. In principle, any material body (or system of bodies – e.g. Earth-Moon system) should respond by displacement to gravitational waves. The stresses of the wave deform the object and the resulting strains indicate and measure the gravitational wave. An elastic body or any physical system sensitive to gravitational waves is called a gravitation antenna. The first gravitation antenna used was an aluminum cylinder, 66 cm in diameter and 153 cm in length. To achieve a good energy transfer from the wave to the antenna, the latter has to be in resonance with the incoming wave. (The fundamental frequency of the mentioned aluminum cylinder is 1661 Hz.) But even in resonance, the deformations produced are very small (of the order of magnitude of Ångstroms and less). They are recorded by special sensors such as piezoelectric crystals, capacitor in a circuit, and laser interferometer which are connected to the antenna. More than a dozen research groups in Europe and America try to detect cosmic gravitational waves. Though much effort has been devoted to the experiments and their interpretation, the results are so far inconclusive.

2.4.7. GRAVITONS

Gravitational fields must also obey quantum laws. Their measurements cannot be absolutely precise, because they have a quantum limit. If this were not the case, the uncertainty principle of Heinsberg would be contradicted. The gravitational theory represents a limiting case for the quantum theory of gravitation (which, however, must yet be constructed). For weak fields the resemblance between the electromagnetic theory and the gravitation theory may be used to construct the quantum theory of gravitation. The quantization may be performed by considering the gravitation in a flat space-time and small corrections to the metric coefficients g_{ik} are considered as field variables. Quantization of the small corrections shows that gravitational waves consist of gravitons.

Gravitons are bosons with spin 2 and the spin projection into the direction of propagation has values +2 and −2. The quantum effects in gravitation are unimportant for astrophysical applications. Therefore one rather speaks about gravitational waves from a close binary and not about gravitons – just as it is common to speak about radio waves rather than radio photons.

Summary

Gravitation is 10^{38} times weaker than the electromagnetic forces and 10^{40} times weaker than the nuclear forces. Nevertheless the gravitational action of a very large number of particles combines together while none of the other interactions can combine as efficiently. On a large scale, gravitation establishes itself as the important force. It plays a fundamental role in the structure and evolution of the universe at large and of its population (satellites, planets, stars, stellar clusters, galaxies, clusters of galaxies).

Among the theories of gravitation two are of eminent importance, viz. the classical Newton theory and the general relativity of Einstein. Both are compared in Table 2.4. The general relativity is one of the metric theories of gravitation which interpret gravitation as curvature of space-time. A series of experiments is planned to find out which of the metric theories explains best the gravitational phenomena in the universe.

General relativity explains all the phenomena that Newton theory of gravitation does. Moreover it explains some observed phenomena that Newton theory could not explain. In general relativity the gravitation is fully described by ten components of the metric tensor. The metric tensor and its derivatives uniquely determine the curvature of space-time. The metric tensor can be determined by solving the Einstein equations. The source of the curvature is given by the mass-energy of all particles and fields.

As a consequence of the field equations, gravitational waves should be emitted by different objects in the universe. Though their observational detection is very difficult and the results of observations still discussed, the gravitational radiation promises to become another important 'window' into the universe.

CHAPTER 3

AGGLOMERATION OF PARTICLES

The number of particles with non-zero rest mass in the observable universe is roughly 10^{80}. All parts of the universe consist of elementary particles, from H atoms to superclusters of galaxies. Each of the particles acts on the other particles and at the same time is exposed to their action. Their interaction agglomerates them into higher units, forms systems of elementary particles and is thus responsible for structure and evolution of the universe.

Not all elementary particles are agglomerated; there are many which move freely through the cosmic space (protons, neutrinos, electrons). But even they are constrained by gravitation and eventually magnetic fields in the space. Each particle in the universe is endowed with an energy (Equation (1.1)) which depends on its rest mass and velocity and has a binding energy (Equations (2.2) and (2.60a)), if it belongs to a system. By interaction with other particles both the total energy (Equation (1.1)) and the binding energies may be decreased or increased.

3.1. States of Matter

The properties of matter (i.e. particle agglomeration) depend on the energy and density of the constituting particles. The dependence of states of matter on the energy and density of their particles is schematically represented in Figure 3.1. If the velocity distribution is maxwellian, temperature may be used on the axis of ordinates.

3.1.1. STATE TRANSITIONS

Many evolutionary processes in the universe represent a transition from one state of matter into another. Thus the abundance of matter in a particular state will change with time. The variety of the states appeared during the evolution of the universe. At its beginning all the particles were in a state represented in Figure 3.1 by points in the upper right corner (fire ball). Expansion of the fire ball, gravitational contraction, strong and electromagnetic interactions distributed the particles in all the other states.

The transitions from one state to another are accompanied by a supply or loss of energy. When plasma is compressed to neutron gas, a large mechanical energy must be applied for the compression. This is done by gravitation at the end of stellar evolution, when a plasmatic star is transformed into a degenerate white dwarf or into a neutron star – depending upon the mass of the collapsing star and hence the pressure applied.

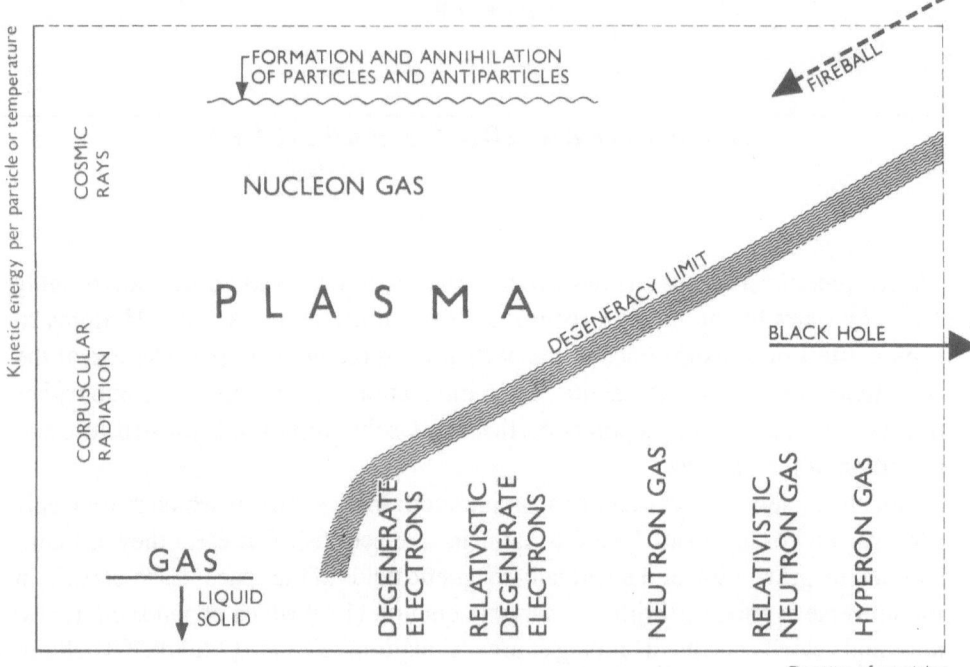

Fig. 3.1. States of matter in the universe. The properties of particle agglomerates depend on particle density and their energy (or temperature). Many processes in the universe are transitions from one state into another. The graph is schematic, units on the coordinate axes are not marked.

By heating a gas it becomes ionized and its properties are substantially changed, so that the matter is called by a different name – plasma. Electrically charged particles of a plasma may gain further energy by acceleration in magnetic fields and they may become high energy cosmic rays. Acceleration of ions by solar flares is an example.

A fine example for the reverse transition may be observed also during solar flares. Corpuscular streamers ejected during solar flares are captured by coronal magnetic fields, condensed and thermalized to plasma state. The hot plasma ($\sim 10^7$ K, $\sim 10^{-14}$ g cm^{-3}) may be observed in coronagraphs or on X-ray photographs of the Sun. When the hot plasma cools to about 10^4 K, spectacular loops are seen through an Hα-filter.

3.1.2. DENSITIES OF MATTER

The range of densities in the universe is very broad, more than forty five orders of magnitude (the mean density of luminous matter is $\sim 10^{-31}$ g cm^{-3}; density of matter in nuclei and in neutron stars is 10^{14} g cm^{-3}). Little is known about the extreme densities in the intergalactic space and in black holes.

The gas densities in interstellar space may vary from 0.1 H atom cm^{-3} to thousands of atoms in some nebulae. The densities of solid state are about 10^{24} to 10^{25}

elementary particles (protons, electrons, neutrons) in one cm^3 (or several $g\,cm^{-3}$). The main contribution to mass of matter comes from the nucleons, electrons representing only a tiny fraction ($\sim 10^{-3}$).

In the interior of stars densities are higher than on our Earth. The material there is compressed by very high pressures not attainable in terrestrial laboratories. The central densities in normal stars depend on their mass and radius; they go from 10^1 to $10^4\,g\,cm^{-3}$. Densities in white dwarfs are $10^5\,g\,cm^{-3}$ and more. In superdense neutron stars the densities are 10^9 to $10^{16}\,g\,cm^{-3}$. The calculated densities in the fire ball were still much higher.

3.1.3. TEMPERATURES AND PARTICLE ENERGIES

The range of temperatures in the universe is also very broad, at least 14 orders of magnitude. The interval of kinetic energies of particles is much broader, about 28 orders of magnitude.

Very low temperatures exist inside dust clouds, from where maser mechanism emits radio lines of OH and H_2O molecules. In the Nuclear Physics Institute in Prague a temperature below 1 mK has been reached by adiabatic demagnetization of diluted cerium-magnesium nitrate. Is this the lowest temperature of the universe?

Very high temperatures exist in stellar plasmas, from a few millions of degrees in young stars to temperatures of $\sim 10^{11}$ K in shock waves formed during gravitational collapse of massive stars. The theoretical temperatures of the fire ball during the first moments of Big Bang should even have been much higher.

The most energetic particles are detected in cosmic rays, viz., protons with energies up to 10^{21} eV. In terrestrial accelerators protons reach energies $\sim 10^{11}$ eV. As may be seen, the intervals of densities and particle energies are much broader in the universe than in terrestrial laboratories.

Our surroundings on the Earth consist of matter in the four states known already to Greek philosophers, solid, liquid, gas and plasma (as fire). The first three – absolutely necessary for life – are rare in the universe, whereas the plasma state, rare on the Earth, is by far the most abundant in the observable universe.

3.2. Solid State in the Universe

Matter in solid state exhibits definite shape and volume. Its constituent ions, molecules and atoms show a regular arrangement, most marked in crystals and present also in amorphous bodies. Interstellar grains, meteorites in the interplanetary space (called also meteoroids), asteroids, planets and their satellites are examples of solid state in the universe. Due to electromagnetic forces, solid bodies resist forces that tend to change their volume or shape. But with increasing mass the self-gravitation increases to the point when the crystallic structure yields (Section 2.4.2). The critical size depends on the material. For rocks it would be about 700 km, for meteorite material the gravitation supersedes the electromagnetic crystalline forces in bodies of about 120 km in diameter. Asteroids probably consist of such a material, so that large asteroids ($\geqslant 120$ km) should be crushed by self-gravitation to a configuration of minimum potential energy, i.e. approximately to a spherical shape. The gravitation is rather inefficient in shaping smaller bodies.

(a) *Planets and Satellites*

The planets and their satellites consist of matter in solid, gaseous and eventually liquid states. Earth, Mars, Mercury, the Moon and some satellites have solid surfaces observable optically. Radar observations of Venus show mountain ranges and large circular basins. It is believed that the interior of Jupiter and Saturn is composed of layers of liquid, solid and metallic hydrogen around a core of dense metallic elements and rocky silicates. The metallic hydrogen in the giant planets may represent 40% of their total mass.

On the Earth, hydrogen is a diatomic gas which becomes solid (still consisting of H_2 molecules) below 14 K. Under extremely high pressure ($>10^6$ atmospheres), hydrogen becomes a metal, with a structure similar to alkali metals. The pressure in the interior of the giant planets reaches 10^8 atmospheres. By high pressures the distance between the diatomic molecules decreases, until an individual atom does not know whether it belongs to its own molecule or the neighboring one. All atoms become equidistant as in a crystal and the hydrogen is in a metallic state. An insulator then becomes a conductor. A similar transition has been observed with phosphorus.

Generally speaking, the properties of solid state depend upon chemical composition and in a complicated way on pressure and temperature. As a result, the interiors of planets are less well known that the interiors of stars, because the properties of stellar plasma are much simpler.

(b) *Meteoritic Complex*

There are many solid bodies in the planetary system, ranging from the size of planets (10^4 to 10^5 km) down to submicron dust particles ($<10^{-4}$ cm). Their material is either earthy like – solid at temperatures up to 2000 K – or ices vaporizing at ≈ 300 K or less, or metallic hydrogen representing a considerable fraction of the giant planets. The total mass of all the solid bodies, i.e. planets, satellites, asteroids, comets, meteorites and dust, equals to 10^{-3} of the solar mass.

The solid state exists also outside the planetary system. Comets (with exception of short-periodic ones) are visitors from the nearby interstellar space. Their solid core (~ 10 km) is surrounded by an icy or snowy volatile envelope. The volatile material is probably similar to a dirty snowball with dust grains. When the comet approaches the interior parts of the planetary system (<3 astronomical units), the volatile envelope sublimates and evaporates. Dragged away by evaporated gases are icy grains, dust grains and dust grains with icy mantle. One type of cometary tails (dust tails) shows only reflected solar spectrum, slightly polarized and reddened. It indicates the presence of fine dust particles, initially dragged away from the envelope by the vaporization gases (300 to 600 m s^{-1}) and then accelerated by solar radiation. The dust emission rate from comets may be a few tons per second, but there are differences between dusty and non-dusty comets.

(c) *Interstellar Dust*

Solid matter exists in space between stars. A considerable fraction of it exists in the form of tiny particles. Solid particles (both dust grains and smoke particles) tend to extinguish the radiation of distant stars – i.e., scatter light to different directions and partly transform the absorbed light into heat. The heat is then reradiated as IR radiation.

The extinction is roughly inversely proportional to the wavelength of the radiation ($\propto \lambda^{-1}$). Longer waves are less influenced so that the light of stars becomes reddened. The regions with increased dust density (dust clouds) are embedded in a general interstellar dust medium. The nearest dust clouds are visible as dark clouds or reflection nebulae if illuminated by nearby stars. The large extinction (~ 1 magnitude per kiloparsec) and the λ^{-1} extinction law imply the existence of grains of size 10^{-5} cm. The space density of the absorbing material is $\sim 10^{-26}$ g cm^{-3}. This means that interstellar dust represents about 1% of the interstellar gas density. Observations of extinction and of the distribution of hydrogen indicate that interstellar dust and gas occur mainly in spiral arms. This is a common feature of all spiral galaxies. In some galaxies the interstellar dust forms a spectacular dark lane in their galactic plane (as in the Sombrero Hat galaxy).

Considerable quantities of interstellar dust are observed near the center of our Galaxy. There are several discrete sources at $100 \, \mu$m coinciding with H II regions and molecular clouds, not far from the center. They are embedded into an extended source at $\lambda \, 100 \, \mu$m, which has dimensions 600 pc $\times 350$ pc and is probably due to a huge cloud of interstellar gas enveloping the central regions of our Galaxy.

It is generally accepted that interstellar grains are formed by condensation of interstellar atoms like smoke and not by the break-up of larger solid bodies like terrestrial dust is formed. But the term 'interstellar dust' is so deeply rooted in astronomers' language that it should not be changed neither to 'interstellar smoke' nor to 'interstellar smog'.

(d) *Solid Intergalactic Matter*

The existence of solid intergalactic matter has been discussed theoretically. Matter in solid state could add an enormous amount of mass to the mean density of the universe, without being detected. The chemical composition of the intergalactic matter should be characterized by a high content of hydrogen; it reflects the abundances of prestellar and pregalactic stage when H (and He) represented all matter in the universe. The solid intergalactic matter may have a form of lumps of hydrogen snow. But it should be stressed that other theories of intergalactic matter suppose that its state is not solid, but rather gaseous or hot plasma or corpuscular radiation. Good observations may resolve the problem.

3.3. Gas in the Universe

The properties of gaseous state are relatively simple. A gas consists of neutral molecules or atoms moving freely, so that they completely fill the space in which they are contained.

3.3.1. ATMOSPHERES OF PLANETS

Neutral gas exists in a relatively narrow temperature interval. Therefore the gaseous state is less often encountered in the universe than the plasma state. Planets possess gaseous envelopes – atmospheres – with chemical composition and physical structure dependent on the planet's mass, its distance from the Sun and on its history. The giant planet Jupiter has sufficient gravity to hold all its atoms including the lightest H. Its atmosphere contains CH_4 and NH_3 besides the abundant H and He.

On the contrary, Mercury has a small mass and a high surface temperature, so that it can retain only the heaviest gases. The atmosphere of Venus consists mainly of CO_2 at its high temperature and pressure, whereas the amount of CO_2 in our atmosphere is substantially smaller – due to photosynthesis of green plants.

The satellite Titan of the distant Saturn possesses an atmosphere consisting of CH_4 and some other gases, among others H and He.

3.3.2. HEADS OF COMETS

Molecules escape from planetary atmospheres into the interplanetary space. But an important supplier of gas to the interplanetary space are comets. The molecules evaporated from the icy nucleus (Section 4.4.3) are dissociated by solar radiation and the resulting neutral atoms or radicals (such as CN, CH, OH, NH, NH_2, C_2, C_3) emit light by resonance excitation. They form a sphere of gas steadily expanding in all directions – coma. The velocities of expansion are slightly less than 1 km s^{-1}, with the exception of H. A huge H coma expands considerably faster, about 8 km s^{-1}, as could be deduced from the Lα isophotes. Gas losses from the coma to interplanetary space may be tens of tons per second.

3.3.3. INTERSTELLAR GAS

Neutral gas forms an important component of the interstellar matter. Its existence has been proved by optical and radio observations. In the spectra of binary stars, the lines of which periodically shift, the interstellar atoms produce sharp stationary absorption lines. In a spectrum of distant stars two or more narrow components are seen, each originating in a separate mass of gas, i.e. in a gas cloud. The cloud structure of the interstellar gas may be observed also in the 21 cm lines of H I. A cloud is probably only a transient gas body of dimensions 7 to 30 pc, changing rapidly and chaotically. It is not held together by self-gravitation.

It is not yet clear how much interstellar gas there is in the space between the clouds. But much gas is concentrated in spiral arms, so that it may be used as a tracer of spiral structure of our Galaxy.

A powerful tool for investigation of physical state and distribution of neutral interstellar gas is the 21 cm line of hydrogen atoms. The integration of the 21 cm profiles over the Galaxy leads to an estimate of 5 to $7 \times 10^9 \, M_\odot$ of H I representing about 4% of the total mass of our Galaxy. This value has been deduced under the supposition that the 21 cm material is everywhere optically thin. Since we now know that a large fraction of interstellar gas is contained in dense and cool clouds with a large optical depth, the estimated amount has to be increased. The densities of interstellar hydrogen have values in a wide range (Table 3.1).

TABLE 3.1

Interstellar densities of hydrogen gas. (Results of different authors summarized by Reddish.)

Object	Number density cm^{-3}
	10^{-2}
Intercloud medium	
	1
Cloudlets	
	10^2
Clouds	
Compact H II regions	
Circumstellar clouds	10^4
Globules	
OH emission sources	10^6
Infrared nebulae	
C 109α emission nebulae	10^8
Ambient nebulae of OH sources	
OH sources	10^{10}

Much neutral gas is concentrated in the central disk of the Galaxy. Neutral molecules OH, H_2CO, CO and others are concentrated in large (30 pc) dense ($\sim 10^3 \, cm^{-3}$) and massive clouds ($\sim 10^6 \, M_\odot$), close to the galactic center (<300 pc). Total mass of molecular gases is guessed at $\sim 10^8 \, M_\odot$, which is more than the total amount of atomic H I observed in the same region.

Neutral hydrogen distribution in some nearby galaxies has been studied by radio telescopes and antenna synthesis. As in our Galaxy, there is a strong contrast between the gas density in arms and that in interarm regions. In spiral galaxies the H I is concentrated in arms and there is a low density in their central regions. The gas component of the galaxies varies with their type: for the elliptical galaxies it is only 0.002 of the total mass, for spirals Sa 0.013, for Sb type 0.03, for Sc galaxies the fraction of the gas mass is 0.2 and in irregular galaxies the gas mass represents a substantial fraction, viz. 0.37.

3.3.4. INTERGALACTIC GAS

The mean density for the closed universe (Section 4.9) should be at least two orders of magnitude higher than the mean density of luminous matter concentrated in galaxies. The intergalactic gas is considered to be one possible form of the missing matter. There is some support for the idea from the 21 cm surveys of nearby galaxies, which show that appreciable quantities of H I exist far outside the luminous galaxies. The motions of such circumgalactic H I indicate that it is bound to the parent galaxy. The gas has apparently too low a density for star formation. It could be still the primordial H. In some cases the gas is observed to fall either from intergalactic space into the galaxy or after a violent ejection from the active nucleus of the galaxy itself (Section 4.7.3).

It is not known whether or not the general intergalactic medium contains any gas. But direct observations indicate that H I exists in intergalactic space near some galaxies (e.g. in the M 81/M 82/NGC 3077 group of galaxies). There are gases ejected by violent galactic activity, infalling high-velocity clouds, uncondensed primordial gas with high angular momentum or intergalactic bridges and galactic tails formed during close encounter of two galaxies.

3.4. Plasma in the Universe

Plasma is ionized gas. It therefore consists of three components, i.e. ions, electrons and neutral atoms. In the case that the ionization is complete, plasma consists only of ion and electron components.

3.4.1. PROPERTIES OF PLASMA

From a microscopic point of view the encounters of plasma particles are governed by electromagnetic interactions between charged particles. Free electrons compensate the charge of positive ions, so that the plasma is neutral. Due to thermal motions of particles, deviations from electrical neutrality are felt in a small region called Debye sphere. Its size is given by

$$\lambda_D = \sqrt{\frac{kT}{4\pi n_e e^2}} = 6.9 \left(\frac{T}{n_e}\right)^{\frac{1}{2}} \tag{3.1}$$

where it may be seen that with increasing temperature the Debye sphere increases. An agglomeration of ions and electrons of size L is called a plasma, if $L > \lambda_D$. Due to deviations from electrical neutrality inside the Debye sphere, plasma is called quasineutral.

If the electron gas in a plasma is shifted relatively to the ion gas, the resulting electrostatic force produces plasma oscillations, in which the potential electrostatic energy of the electrons is transformed into kinetic energy and vice versa, with a plasma frequency

$$\omega_p = \sqrt{\frac{4\pi n_e e^2}{m_e}} \approx 5 \times 10^4 \sqrt{n_e}. \tag{3.2}$$

This important plasma characteristic depends only on electron density. To incident electromagnetic waves with frequency ω a plasma appears as an insulator, if $\omega_p < \omega$ and they pass through the plasma. If $\omega_p > \omega$, the waves are absorbed and do not propagate.

3.4.2. MAGNETIZED PLASMA

Plasma is a good conductor of electricity. In an electric field its electrons and ions are accelerated in reverse directions and produce electric current. The plasma follows the law of electrodynamics (Maxwell equations and Ohm law). Plasmas also obey the laws of hydrodynamics. The behavior of plasmas in magnetic fields is studied by hydromagnetics (called also magnetohydrodynamics, cosmical electrodynamics). Hydromagnetics uses laws of both electromagnetics and hydrodynamics. The coupling of both is by the Lorentz force, which is the force acting on a charge and current in a magnetic field. The hydromagnetics proved very useful in understanding many phenomena in the universe. Plasma is almost ubiquitous and magnetic fields permeate the whole universe.

The Lorentz force acting on an electron or ion moving with velocity $\mathbf{v}$ in a magnetic field $\mathbf{B}$ is

$$\frac{q}{c}(\mathbf{v} \times \mathbf{B}), \tag{3.3}$$

where q is the charge of the particle (for electron $-e$, for proton $+e$, for α-particle $+2e$, etc.). The cross product of both vectors means that the Lorentz force is perpendicular to both. It means that the charged plasma particle have to spiral around lines of force (Figure 3.2). It is convenient to regard magnetic lines of force as if they had a real existence. The circular frequency (angular velocity, gyrofrequency, Larmor frequency) ω_g

$$\omega_g = \frac{qB}{mc} \tag{3.4}$$

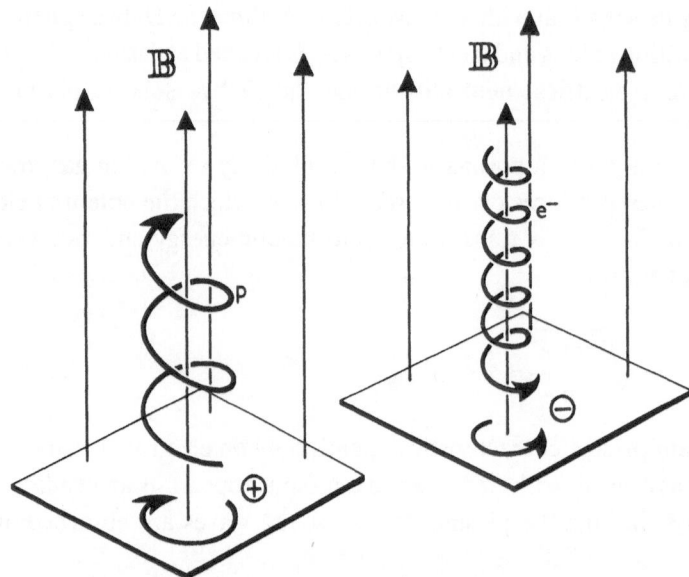

Fig. 3.2.   Motions of ions and electrons in a magnetic field. By the motion particles are bound to lines of force; in other words 'field is frozen into the plasma'. The smooth spiral motions are disturbed by collisions, in particular in dense plasmas.

so that it is independent from the velocity of the particle. For protons the gyrofrequency is approximately $10^4\ B$ when B is expressed in G and gyrofrequency in Hz. For electrons the corresponding value would be $1.7 \times 10^7\ B$. The radius of gyration (gyroradius, Larmor radius) of the gyrating particle

$$r_g = \frac{mcv_\perp}{qB} \tag{3.5}$$

is proportional to the velocity component $v_\perp$ perpendicular to the magnetic field.

As a consequence of the particle gyration described by Equations (3.4) and (3.5), the plasma and magnetic field are coupled together. The coupling is usually described symbolically as 'the lines of force are frozen into the plasma'. Plasma with a frozen magnetic field is called magnetized plasma. A magnetized plasma constitutes an anisotropic medium.

The degree of coupling between the plasma and the magnetic field depends on the plasma conductivity. As the conductivity is finite, the magnetic field decays with time and transforms its energy into Joule heat. Another possibility to change the magnetic fields in the plasma is to compress it, because plasma is compressible like a gas. The change of magnetic fields with time is implicitly given by the Maxwell equations or explicitly by an equation which follows straight from them:

$$\frac{\partial \mathbf{B}}{\partial t} = \nabla \times (\mathbf{v} \times \mathbf{B}) + \frac{c^2}{4\pi\lambda} \nabla^2 \mathbf{B} \tag{3.6}$$

where λ is the conductivity of the plasma, $\mathbf{v}$ its velocity and ∇ the delta operator.

For a plasma with low electrical conductivity and in slow motion ($\mathbf{v} \sim 0$), the first term on the right hand side may be neglected. The resulting equation

$$\frac{\partial \mathbf{B}}{\partial t} = \frac{c^2}{4\pi\lambda} \nabla^2 \mathbf{B} \tag{3.7}$$

is a diffusion equation for the magnetic field in plasma.

It may be approximated by an algebraic equation

$$\frac{B}{t_0} = \frac{c^2}{4\pi\lambda} \frac{B}{L^2}, \tag{3.7a}$$

where L – the characteristic dimension of the plasma stands for the operator ∇. For the decay time t_0 we get

$$t_0 = \frac{4\pi\lambda}{c^2} L^2. \tag{3.8}$$

It is worth noting the strong dependence on the characteristic dimensions of the plasma cosmic body. This is the reason why magnetic fields are frozen into large interstellar clouds, even if their ionization and conductivity are very low.

If, however, the conductivity is large and velocities of the plasma not small, then the second term of Equation (3.6) may be neglected and the resulting equation expresses in mathematical form that the lines of force are frozen into the plasma. Due to this coupling (expressed microscopically by Larmor gyration reactions (3.4) and (3.5)), the fate of plasma is closely connected with the fate of the magnetic field. Thus plasma motions shape and intensify the frozen-in fields, or vice versa magnetic fields influence the dynamics of the frozen-in plasmas. It depends on which of the two partners is stronger in a particular case. This may be found out by comparing the kinetic energy of the plasma $\frac{1}{2}\varrho v^2$ with the energy density of the field $B^2/8\pi$:

$$\frac{1}{2}\varrho v^2 < \frac{B^2}{8\pi} \qquad \text{or} \qquad \frac{1}{2}\varrho v^2 > \frac{B^2}{8\pi}. \tag{3.9}$$

Quiescent prominences on the Sun are an example of the second inequality; their weak fields are easily twisted by the chaotic motions of the prominence plasma (Figure 3.3). On the contrary, the smooth regular streamers of plasma in coronal loops indicate strong magnetic fields over sunspots; in the loops the first inequality (Equations (3.9)) is valid (Figure 3.4). Generally speaking, interaction of plasma with magnetic fields produces many phenomena in the universe, often of unusual beauty: aurorae, cometary tails, eruptive prominences, remnants of supernovae etc. It produces also all phenomena of solar activity (plages, sunspots, flares, coronal condensations, prominences) with the consequences for the Earth and its biosphere.

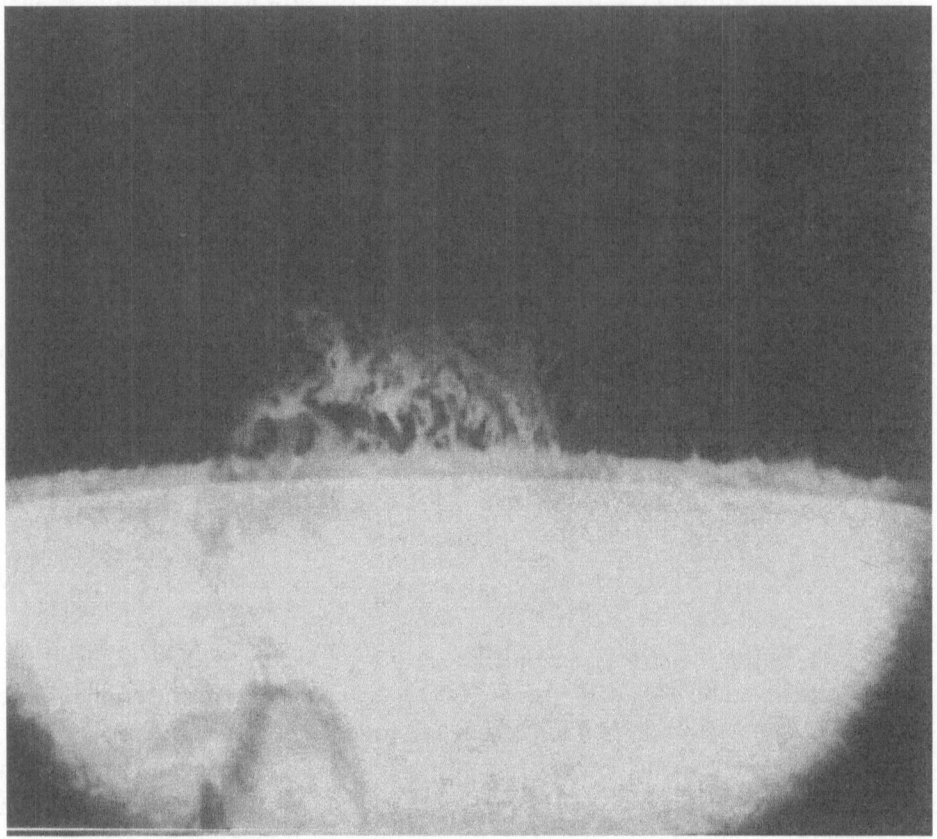

Fig. 3.3.   A quiescent prominence with chaotic motions. (Hα photograph of Sacramento Peak Observatory, AFCRL, 1970, July 29). Energy density of plasma prevails over the magnetic energy (Second
inequality 3.9).

3.4.3. WAVES IN MAGNETIZED PLASMAS

A magnetized plasma transmits a variety of waves. The energy contained in a cm^3 of
plasma is the sum of electrostatic energy, magnetic energy, kinetic energy of
electrons and ions, thermal kinetic energy of the particles, ionization and excitation
energy and gravitational potential energy. One form of energy may be periodically
transformed into another one and vice versa, producing a type of plasma wave.

The various modes of plasma waves may be reduced to three basic types: (a) The
electrostatic interaction between the ions and electrons give rise to two longitudinal
modes of electrostatic waves. In one mode both components move together and form
a sound wave. In the other mode of electrostatic waves – called plasma oscillations –
the ions and electrons oscillate 180° out of phase. (b) The electromagnetic waves are
transverse and are distinguished by their polarization with respect to the magnetic
field of plasma. (c) Another type of transverse wave – hydromagnetic waves – may
propagate only in a magnetized plasma. The frozen-in field exhibits the restoring

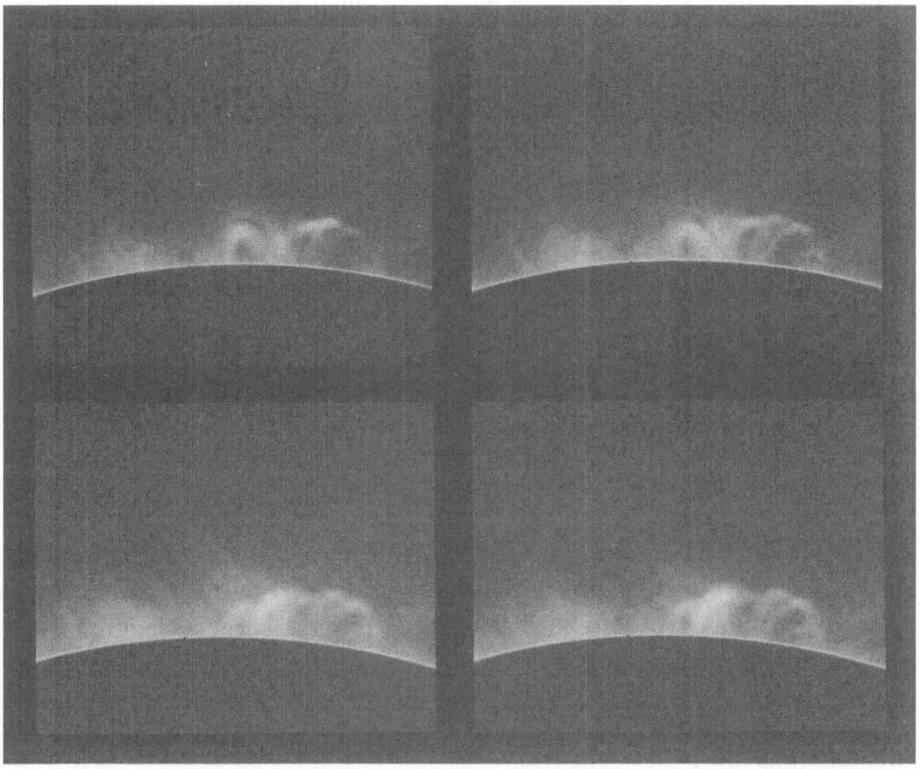

Fig. 3.4. Coronal plasma in strong magnetic fields above an active region. The behaviour of the plasma is fully controlled by the strong magnetic fields (3.9 first inequality). Photograph in the light of 5303 Å of Fe XIV (Sacramento Peak Observatory).

force against shearing motions of plasma. Such forces exist neither in a plasma without magnetic fields nor in a gas. The best known example of hydromagnetic waves is the Alfvén waves propagating with velocity

$$v_A = \frac{B}{\sqrt{4\pi\varrho}}.$$ (3.9a)

Another example are the collisionless shocks which cannot be studied in terrestrial laboratories, but are produced in interplanetary space and studied *in situ*. The gyration radius of ions replaces their mean free path (which is many orders of magnitude longer). The thickness of the collisionless shock is a few gyration radii and the only collisions are due to magnetic fields.

A disturbance in a plasma may propagate as a wave with periodic transformation of energy, such as electrostatic⇄kinetic (plasma oscillations), electric ⇄magnetic (electromagnetic wave), kinetic⇄magnetic (Alfvén wave) etc. However, the structure of a plasma may be such that an initial disturbance steadily grows so that the transformation proceeds in one direction only and the plasma gets rid of some energy to become more stable. Such transitions are called plasma instabilities or hydromagnetic instabilities. Different types of the instabilities are observed in the universe and also in plasmas produced in terrestrial laboratories. Useful to astronomers for interpreting many astrophysical processes, the instabilities plague the plasma physicists, preventing the production of thermonuclear reactions on our planet. There are many institutes where plasma is produced, studied and applied for practical purposes (chemical synthesis, growth of crystals, cutting of metals, radiation sources, etc.).

3.4.4. IONOSPHERES

Plasma constitutes almost the entire material content of the observable universe. The nearest plasma is the ionosphere, which extends from about 50 km upwards. The upper layers of our Earth (and also of other planets) are a complicated plasma system, which is ionized by short-wave radiation in a broad spectrum from UV to X-rays and also by cosmic rays and electrons precipitated from the magnetosphere. For example the atmospheric molecules N_2 are ionized by solar radiation with wavelengths shorter than 769 Å (including the strong He I line 584 Å). The radiation required for ionization of atomic oxygen has wavelengths below 911 Å while for the O_2 ionization $\lambda < 1027$ Å are necessary. The strong Lyman lines maintain ionization in altitudes 90 to 150 km (the so called E region of the ionosphere). But the X-rays in the range 8 to 140 Å are about equally important at these altitudes.

At longer UV wavelengths the photons may penetrate deeper, but they do not possess enough energy to ionize the main atmospheric constituents. For the D region in altitudes 60 to 90 km the $L\alpha$ (1216 Å) and the X-rays (1 to 10 Å) are the main source of ionization. In the lowest part of the ionosphere (50 to 60 km, the C-region) cosmic rays contribute to ionization. The corpuscular radiation plays a role in ionization during aurorae. Generally speaking, the ionosphere is a complicated plasma system with numerous ions, subject to ionization, losses by recombination and transport processes that vary with altitude, geographical position, with local time and with the solar activity. The ionosphere is of fundamental importance for the biosphere by absorbing the dangerous X-rays and UV solar radiation. Because of its property to reflect short radio waves it is of basic importance for radio communication all over the world.

We have direct observational evidence for ionosphere on Venus and Mars. On Venus the ionosphere extends from ~80 km to ~200 km, with maximum electron densities 10^5 electrons in cm^3 at the altitude of 130 km. Moon is also surrounded by a layer of electrons, which of course do not form a plasma (Section 3.4.1). The electrons are emitted by photoeffect from the illuminated lunar soil. Theoretical considerations indicate that observable ionospheres should exist on Jupiter, Saturn, Uranus and Neptune. Protons should represent the dominant plasma ions, due to the high abundancy of hydrogen. Radio measurements of the occultation of Pioneer 10 by the satellite Io has even proved existence of the ionosphere on the satellite, extending to about 700 km above its surface, with a peak electron density of $6 \times 10^4 \, \text{cm}^{-3}$, at altitudes 60 to 140 km.

3.4.5. INTERPLANETARY PLASMA

The interplanetary space is filled with expanding interplanetary plasma, which contains also a weak magnetic field of the order of a few gammas (1γ is 10^{-5} G). The field has been measured by instruments on several interplanetary space probes; it is highly variable in intensity and also in direction.

Nuclei of comets eject plasma into the interplanetary space. The cometary ions (e.g. CO^+, CO_2^+, N_2^+) are accelerated from the nuclear region by solar wind and form a long, straight tail with fine structures, such as nodules, twists, etc. The spectra of plasma tails show bands of molecular emission in contrast to dust tails which reflect solar spectrum.

While comets are only temporary suppliers of some interplanetary plasma, the Sun is a powerful source of a steady flow of plasma in the form of the solar wind. It flows away from the Sun with a supersonic and superalfvénic velocity ($>B/\sqrt{4\pi\varrho}$). Its properties are summarized in Table 3.2. The kinetic energy of the solar wind is about two orders of magnitude larger than its magnetic energy. As a consequence of the inequality (second inequality in formula (3.9)), the magnetic fields are carried with the solar wind away from the Sun, without braking it.

TABLE 3.2

Typical properties of the solar wind
during quiet periods (Hundhausen)

Solar wind velocity	$320 \, \text{km s}^{-1}$
Proton density	$8 \, \text{cm}^{-3}$
Total mass loss of the Sun by the solar wind	$1.2 \times 10^{12} \, \text{g s}^{-1}$
Proton temperature	$4 \times 10^4 \, \text{K}$
Electron temperature	$1 - 1.5 \times 10^5 \, \text{K}$
Magnetic field strength	$5 \times 10^{-5} \, \text{G}$
Alfvén velocity	$38 \, \text{km s}^{-1}$

Solar wind consists mainly of H and its He component varies from a few percent to about 20%. Heavier elements have also been identified; however, the solar wind contains not only plasma with temperatures of about 10^5 K. The particle detectors on interplanetary probes measure suprathermal particles up to cosmic rays. Though the energies of the energetic particles are high, their energy density is much less than that of the magnetic fields. Hence the hierarchy is such that the motion of solar wind controls the frozen-in magnetic fields and the magnetic fields control the motion of cosmic ray particles.

The solar wind is a complex phenomenon. Its velocities and densities are variable and depend on solar activity. It is inhomogeneous in density and in magnetic field, it contains large-scale plasma streamers, small-scale filamentary structures, different types of discontinuities (such as collisionless shock waves or a standing shock wave above the Earth's magnetosphere) or small scale variations (such as Alfvén waves with wavelengths 10^3 to 10^6 km which propagate outward from the Sun and have large amplitudes of several gammas).

Large plasma clouds are ejected from the solar atmosphere during active periods. Plasma masses up to 10^{14} g are observed to be ejected in the form of eruptive prominences, sprays, ascending filaments and expanding coronal structures. Flares are especially powerful sources of ejected plasma clouds.

3.4.6. SOLAR AND STELLAR PLASMA

Our Sun is a plasma sphere and its mass $(M_\odot = 2 \times 10^{33}$ g$)$ is three orders of magnitude larger than the mass of all the other members of the planetary system together. This relative abundance of plasma in the planetary system seems to be roughly the same in the whole universe. But even the matter of cool planets, comets, of all the members of the meteoritic complex, of the whole biosphere including our own bodies – the atoms of all the matter passed through the plasma state at least once during the history of the universe. In its pregalactic phase all matter existed in the form of H (and He) plasma. A considerable fraction of (if not all) cosmic matter passes sometimes through the stellar phase; it is enriched by heavy elements in the hot stellar interior to the detriment of H II. The mean atomic number of cosmic material thus increases and the mass defect ($\sim 10^{-3}$ of rest mass) is the energy source for stars.

By far the greatest part of our Galaxy consists of stellar plasma. The interstellar dust represents only a few percent of the mass of the interstellar matter, which is about 10^{-3} of the mass of the Galaxy.

The interstellar gas around hot O and B stars is ionized and forms H II regions (called sometimes Strömgren regions or Strömgren spheres). The strong output of UV radiation from these stars maintains the H in H II regions in a state of ionization. Recombination of electrons with protons emits radio lines and also a strong Hα line, in which the H II regions are observed.

The third component of the interstellar matter, the neutral gas far from hot stars is also slightly ionized by interstellar photons and by cosmic rays. Mainly the atoms with low ionization potential are ionized and there is observational evidence for free electrons in neutral interstellar gas (0.03 cm^{-3} averaged over arm and interarm regions). The value of the electron density is provided by dispersion of radiation from pulsars, the distances of which are known (e.g. through the λ 21 cm absorption). The dispersion means that pulses on different wavelengths suffer different time delay when traversing through the interstellar matter.

Even the interstellar gas the H component of which is neutral, is a plasma, with very small ionization degree $(10^{-3}-10^{-4})$. The magnetic fields remain frozen in it due to large dimension L in formula (3.8).

We may say that practically all the observable mass of the Galaxy (about 99.9%) is in the plasma state. Interstellar dust represents only a small fraction, like solid state in the solar system, i.e., about 0.1%. Moreover, the dust grains also have electric charge and in some respect behave like ions: they circle with gyrofrequency (Equation (3.4)) in interstellar magnetic fields, may be accelerated to high energies etc. The importance of electromagnetic interactions everywhere in our Galaxy thus becomes evident.

3.5. Corpuscular Radiation

Most molecules, atoms, electrons and ions in cosmic space have thermal energies ranging from $\sim 10^{-3}$ to ~ 1 eV. Relatively very few interplanetary and interstellar, planetary and solar particles have suprathermal energies. They move faster than thermal particles in the surrounding plasma. A general name 'corpuscular radiation' is used for them, to differentiate them from plasmas with maxwellian distribution of velocities. However, the distinction is not always clear; a cloud of charged particles ejected from solar atmosphere is a plasma cloud for a co-moving observer, because the velocity distribution of the particles is for him maxwellian. On the contrary, an observer in a spaceship will measure velocities one to two orders of magnitude larger (i.e. 10^2 to 10^3 km s^{-1}) than the thermal velocities inside the cloud (about 10 km s^{-1}). The solar wind is sometimes considered to represent an expanding coronal plasma and at other times it is held for corpuscular radiation. But even in the first concept of the solar wind, there are suprathermal particles ranging up to $\sim 10^{10}$ eV.

The corpuscular radiation incident on our Earth has a very broad energy spectrum, ranging from the solar wind particles with energies 10^2 to 10^4 eV up to energies of at least 10^{20} eV of some particles coming from galactic or intergalactic space. The most abundant ions in the incident corpuscular radiation are protons which reflect the common composition of plasma in the universe (Figure 4.3). They interact with the terrestrial atmosphere and produce many secondary particles (Figure 3.5).

The concentration of high energies into some particles of cosmic rays seems to be against the tendency of nature, because it means decrease of entropy. If the second principle of thermodynamics is not to be violated, the decrease of entropy in accelerated particles must be correspondingly compensated by an entropy increase in the accelerating system. At present there is no unique theory to explain the acceleration to high energies either of local (magnetospheric, solar) or of galactic ions. Variable magnetic fields will probably play an important role in many accelerating processes in the universe.

3.5.1. Radiation belts

The terrestrial magnetosphere is filled with ions and electrons. Two regions of enhanced corpuscular radiation were recognized, with centers having a distance of

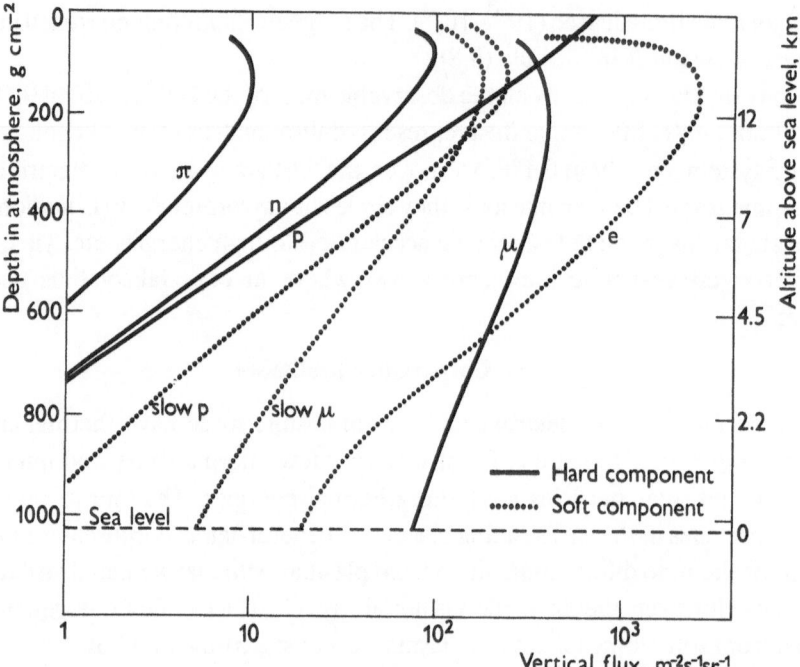

Fig. 3.5. Corpuscular radiation in the terrestrial atmosphere (cosmic radiation). The hard component particles π, n, p and μ have high energies ($>4 \times 10^8$ eV, $>3 \times 10^9$ eV, $>3 \times 10^9$ eV, and $>2 \times 10^8$ eV). The soft component particles p, μ, e have low energies (0.4–3×10^9 eV, 0.27–2.2×10^8 eV and $>10^7$ eV). (After Peters.)

1.5 R_E and 3.5 R_E from the Earth's center and above the equator. The ionospheric plasma extending to about 1.3 R_E is dense enough to be collision dominated and currents may therefore flow along, as well as across, the lines of fields. The collisions of the electrons and ions in the radiation belts are rare and their motion is governed mainly by magnetic fields.

Within the magnetosphere the electrons range from thermal energies ($\sim$eV) up to several MeV. The proton energies extend from thermal energies to several hundreds of MeV. In the inner belt protons with energies up to 10^3 MeV constitute the principal kind of high-energy particles. It is generally accepted that their source is the atmospheric neutron albedo (i.e., neutrons resulting from collisions of cosmic rays with nuclei of atmospheric gases). The neutrons move from the lower atmospheric layers across the ionosphere and magnetic fields upwards. They decay (Equation (2.19)) into protons which are captured in the magnetosphere.

The outer belt is a region of energetic electrons (up to a few MeV) with the core at $\sim$16 000 km above the Earth's surface. It is an extremely tenuous medium, with a total mass of about 15 kg. The protons in the outer belt possess much lower energies than those in the inner belt. The electrons undergo large variations in number and energy, especially during magnetic storms.

The radiation belts of the Earth are not unique in the planetary system. Non-thermal radio emission from relativistic electrons in Jupiter's magnetosphere indicates intense fluxes of high-energy particles around this giant planet. Recently the fluxes have been measured *in situ* by instruments aboard the Pioneer spacecraft. For large distances from the planet (i.e., 30 to 100 R_J) high energy electrons and protons are concentrated near the magnetic equatorial plane. For small distances (< 20 radii) the behavior of radiation is similar to the terrestrial radiation belts. Maximum measured fluxes in the Jupiter's radiation belts were 4×10^6 protons cm^{-2} s^{-1} (for energies >30 MeV) and 5×10^8 electrons cm^{-2} s^{-1} (for energies >3 MeV).

3.5.2. SOLAR CORPUSCULAR RADIATION

The most powerful emitter of energetic particles in the solar system is the Sun itself. Besides the solar wind (Figure 3.6), the cosmic and subcosmic rays (with energies $>10^9$ eV or $<10^9$ eV) are emitted during discrete events. Some events are barely detectable, with fluxes of about one particle cm^{-2} sr^{-1} s^{-1} (for energies >1 MeV) and no observable association with flares or other optical phenomenon. On the other side, there are flare-associated events with maximum energy over 10^{10} eV and fluxes over 10^3 protons cm^{-2} sr^{-1} s^{-1} for more than a day (for energies >20 MeV).

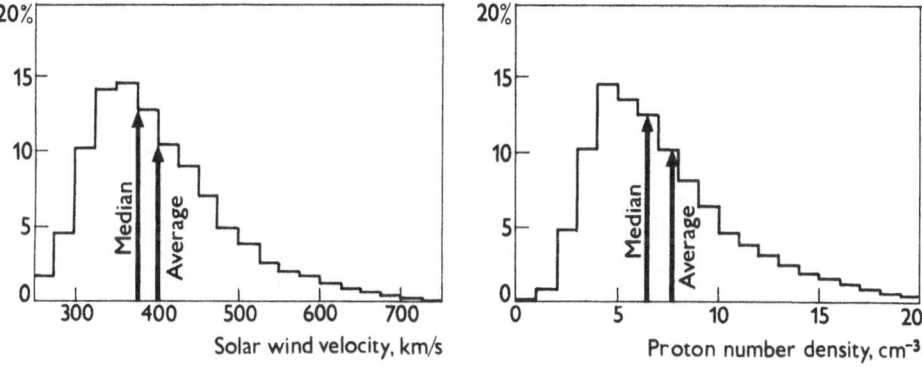

Fig. 3.6.   Solar wind measurements: distribution of velocities and energies (Fälthammar).

The total energy (about 10^4 erg cm^{-2} sr^{-1}) for the largest solar cosmic ray event arriving at the top of our atmosphere is comparable with the energy brought by the galactic cosmic rays in 1 yr. The spectra of the solar particle events are much steeper than the spectra of cosmic rays.

Protons are the predominant particles in the solar wind but other nuclei are also detected. Electrons are also recorded in solar cosmic-ray events by balloons, satellites and interplanetary probes. During some electron events no protons have been observed.

The high energy particles are mainly accelerated in the solar atmosphere above active regions with strong magnetic fields. Though details of the acceleration process are not fully explained, the changing magnetic fields play a fundamental role in accelerating thermal ions and electrons in the solar atmosphere to relativistic energies of cosmic rays (Figure 3.7). Both are observed in the solar atmosphere. Electrons radiate more efficiently than ions do, due to their small mass and large charge/mass ratio. They produce X-ray and radio bursts during flares. Energetic protons produce spallation reactions in the solar atmosphere, leading to γ-ray line emission. Detectable γ-ray fluxes are observed from the most powerful flares.

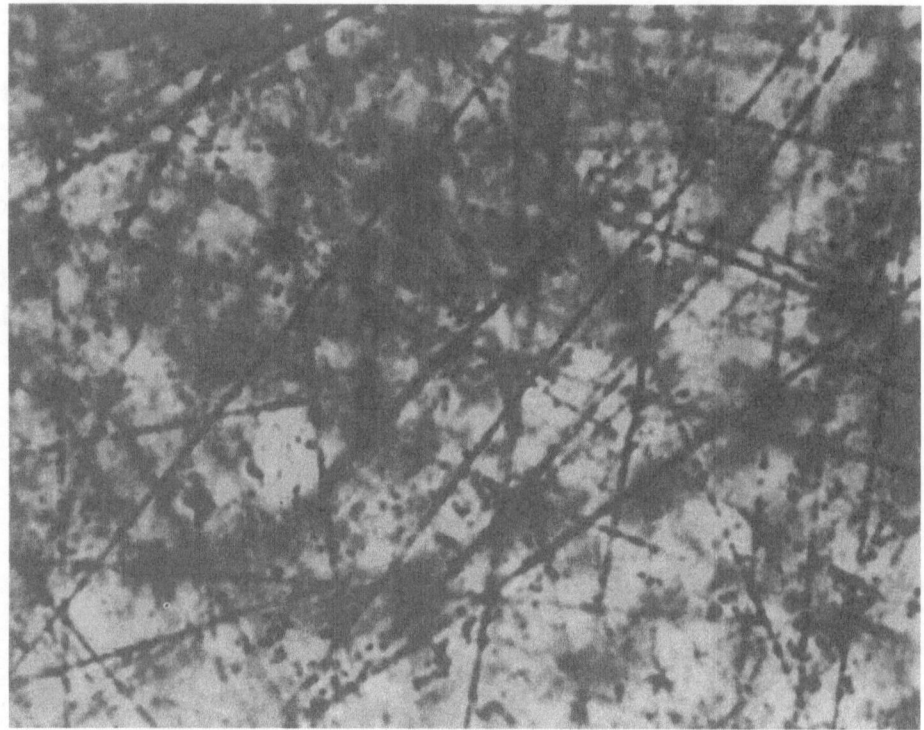

Fig. 3.7.   Tracks of relativistic solar particles (protons) recorded after a flare (1960, November 12). The nuclear emulsion was brought above the atmosphere by a rocket (Goddard Space Flight Centre).

3.5.3. Galactic cosmic rays

Recent radio observations indicate that cosmic rays are omnipresent not only in the Galaxy, but also in other galaxies. The synchrotron radiation on decimeter and meter wave-lengths is emitted by relativistic electrons spiralling in magnetic fields. The relativistic electrons manifest their existence still in another way, viz., by producing γ-photons. There are two possible mechanisms: (a) bremsstrahlung of high energy electrons when they encounter an ion, and (b) inverse Compton effect when a high energy electron transfers its energy to an optical photon.

The mechanism which accelerates electrons to relativistic energies accelerates also positive ions. As a result, relativistic protons and other nuclei exist in interstellar space, but their synchrotron emission is much weaker than the emission of relativistic electrons. Since the charge-to-mass ratio is three orders of magnitude smaller for ions than for electrons, this acceleration is also smaller by the same factor.

Although not observed in synchrotron radiation, the cosmic-ray nuclei contribute to the γ-radiation of the Galaxy. In collisions with interstellar nuclei many new particles and anti-particles are produced, including neutral pions. The pions decay after about 10^{-15} s into two photons (Equation (2.4)). Each of the three abovementioned γ-emitting processes has its own characteristic spectrum and may thus be identified.

The energy contained in galactic cosmic rays ($\sim 10^{-12}$ erg cm^{-3}) is comparable to the energy density of magnetic fields of the Galaxy ($B^2/8\pi = 10^{-12}$ erg cm^{-3} for 5×10^{-6} G), to the kinetic energy of interstellar gas ($\frac{1}{2}\varrho v^2 = 0.8 \times 10^{-12}$ erg cm^{-3} for two protons in a cm^3 and $v \approx 7$ km s^{-1}) and to total star light (0.7×10^{-12} erg cm^{-3}). Cosmic rays (Figure 3.8) thus represent an important dynamical factor for our

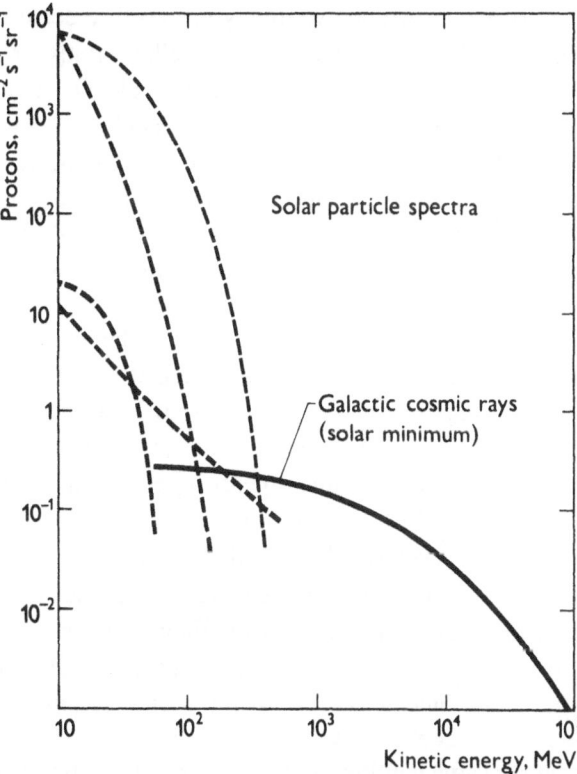

Fig. 3.8. High energy protons in galactic cosmic rays (full curve). Protons emitted from different solar flares are represented by dashed lines. While the flux of galactic cosmic rays remains constant (slight variations are due to interplanetary magnetic fields), the solar protons show a strongly variable distribution.

Galaxy and for the other galaxies as well. The equality of energy densities corresponds to an approximate equality of pressures ($erg\,cm^{-3} = g\,cm^2\,s^{-2}\,cm^{-3} = dyn\,cm^{-2}$). The equality of the cosmic ray pressure with magnetic field pressure may be interpreted as an accumulation of cosmic rays in the galactic magnetic fields, until their pressure suffices for them to escape into the intergalactic space.

The cosmic ray particles undergo many alterations on their journey through interstellar space. They are deflected by magnetic fields and their inhomogeneities. Thus, for example, the deflection by interplanetary magnetic fields prevents the particles with lower energies from arriving at our Earth. The interplanetary field is a consequence of solar activity; it changes with the solar cycle and so does the number of galactic cosmic ray nuclei with energies $\lesssim 10^{10}$ eV. This phenomenon is called the modulation of galactic cosmic rays.

The deflections of cosmic rays by local magnetic inhomogeneities and by collisions with interstellar plasma clouds randomize the velocity vectors – thus producing an isotropy of arrival directions upon the Earth's magnetosphere. Another electromagnetic effect on cosmic-ray nuclei is the ionization loss: by traversing the interstellar gas, the nuclei ionize its atoms and lose energy. The ionization losses are pronounced for heavy ions and for lower energies.

Deeper alterations in energy and eventually in structure occur if a heavy cosmic ray nucleus collides with the nucleus of interstellar atoms or ions. By strong interactions heavy nuclei are fragmented in light nuclei like D, Li, Be, B, F (spallation reactions). Thus the abundance of light nuclei in cosmic rays is increased (Section 2.1.4). The flux of the light nuclei incident on our Earth will therefore depend on the number of collisions of the heavy parent nuclei, i.e., on the mean age of cosmic rays. It turns out that on the average, a cosmic ray nucleus should live only for a few millions of years. This is a relatively short time – about three orders of magnitude shorter than the life of a normal star.

On the other hand, the cosmic–ray archeology of the planetary system supports the idea that the galactic cosmic-ray flux has not changed appreciably during the last $\sim 10^8$ yr. (The fluxes during the past are deduced from the abundances of cosmogenic nuclei produced by cosmic rays in solid state bodies, such as meteorites, or rocks on the lunar surface.) It follows that cosmic rays are a permanent feature of the Galaxy and some sources have to exist to replenish them.

Interstellar ions may be accelerated by elastic collisions with clouds of magnetized plasma or by colliding hydromagnetic shock waves (Fermi mechanism). But discrete sources seem to be more efficient accelerators. Electrons are accelerated in supernova remnants to relativistic energies, and it is highly probable that ions also are accelerated by the same mechanism. With the identification of neutron stars in supernova remnants (NP 0532 in Crab Nebula and PSR 0833-45 in the Gum Nebula), the pulsars became leading candidates for acceleration of cosmic rays. The observed secular increase of their rotational period (i.e., decrease of rotational energy of the neutron stars) is in agreement with the idea that pulsars are cosmic-ray sources.

3.6. Matter at High Temperature

Plasma contains not only ions and electrons but also photons. The energy density of photons ε_γ in a plasma increases fast with its temperature:

$$\varepsilon_\gamma = aT^4 \tag{3.10}$$

where the radiation constant $a = 7.56 \times 10^{-15}$ erg cm^{-3} deg^{-4}. The γ-photons in a very hot plasma ($\sim 10^{10}$ K) dissociate atomic nuclei. The process of photodissociation affects even the most stable nuclei, dissociating them into α-particles and eventually into neutrons and protons:

$$^{56}\text{Fe} \rightarrow 13\,^4\text{He} + 4\,\text{n} \tag{3.11}$$

and

$$^4\text{He} \rightarrow 2\,\text{p} + 2\,\text{n}. \tag{3.12}$$

Thus neutrons appear in very hot plasmas.

An important property of a very hot plasma is the creation of particles, not only of photons. Neutrinos are formed in pairs with antineutrinos (Equations (2.32a) to (2.41)). They have zero rest mass and their energy density is close to the same quantity for photons. In the last phases of evolution of massive stars the neutrino-antineutrino formation may represent an efficient cooling mechanism for the hot stellar interior.

At temperatures 10^{10} K, corresponding to $2\,m_e c^2 = 10^6$ eV (1 eV corresponds to 10^4 K), electron-positron pairs are materialized (Equation (1.17)). At still higher temperature ($\sim 10^{12}$ K) mesons and antimesons are created; further nucleons, antinucleons (p, p̃, n, ñ) and hyperons at temperatures $\sim 10^{13}$ K appear in the plasma. The role of particles with mass m is relatively small – of the order $e^{-mc^2/kT}$ – if $mc^2 > kT$. On the contrary, at very high temperatures $kT > mc^2$ the reactions of materialization occur at a very high rate so that all elementary particles are in thermodynamical equilibrium. The statistics give for their energy densities:

$$\varepsilon = \left(\frac{g}{2}\right)\varepsilon_\gamma \text{ for bosons} \tag{3.13}$$

and

$$\varepsilon = \left(\frac{7}{16}\right)g\varepsilon_\gamma \text{ for fermions.} \tag{3.14}$$

In both expressions g means statistical weight of the particles ($g_\gamma = 2$, $g_\pi = 3$, $g_e = 2$, $g_\mu = 2$, $g_\nu = 1$ etc). According to Equation (3.10) ε_γ is the energy density of photons.

There remains, of course, a question whether plasmas with such high temperatures exist at present anywhere in the universe. In the past, Big Bang may have provided a short opportunity for such a prodigious formation of all possible elementary particles and antiparticles (Section 5.2).

3.7. Matter at High Densities

A point in Figure 3.1 represents temperature and density of matter. Heating shifts the point upwards, while compression increases number of particles in a unit volume and the representative point shifts to the right. For simplicity one considers matter at relatively low temperatures, called cold matter; this term applies to matter with temperatures not sufficiently high to change substantially its properties, i.e. not higher than the degeneracy temperature (defined later). For example a matter with $10^6 \, \mathrm{g \, cm}^{-3}$ and $10^8 \, \mathrm{K}$ is a cold matter. The importance of cold dense matter in astrophysics has been stressed by Ambartsumian. In 1960 he and Saakyan gave the first detailed analysis of very dense matter.

3.7.1. MATTER AT LOW DENSITIES

At low densities and temperatures ($\varrho < 50 \, \mathrm{g \, cm}^{-3}$ or pressure 5×10^6 atmospheres) the matter appears in a great variety of forms. Its chemical and physical properties are sharply defined and change from one element to another in a non-monotonous way. Pressures up to 10^7 atmospheres have been produced in laboratories by explosions, to study properties of individual elements. In simple cases properties may be deduced theoretically for a given density and temperature. This is in the case of hydrogen.

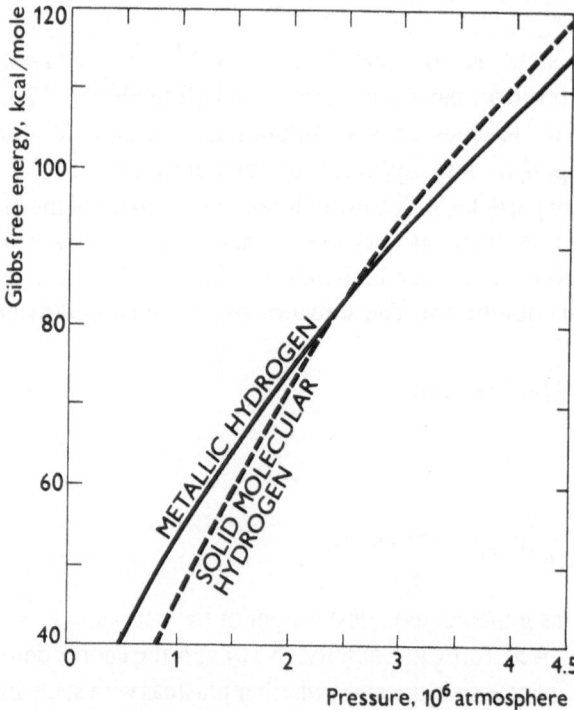

Fig. 3.9. Gibbs free energy for hydrogen in solid molecular state and in metallic state. (Free energy of a system is that portion of its energy which may be converted to work). For pressures $\geqslant 2.5 \times 10^6$ energetically more stable is the metallic state.

If solid H_2 is compressed by a pressure of about 2.5×10^6 atmospheres, the atoms are ordered to a crystal lattice characteristic for alkali metals, with 5×10^{22} atoms per cm^3. For still higher pressures and densities the solid H_2 would have considerably higher free energy than the metallic hydrogen, as may be seen from Figure 3.9. Gibbs free energy $U - TS + pV$ is the energy of the matter which may be changed into work (U is the internal energy, T, temperature, S, entropy, p, pressure, V, volume). In recent laboratory experiments (by Grigorev) a compressed H_2 sample produced a discontinuity from 1.08 g cm^{-3} to 1.3 g cm^{-3} at a pressure of about 2.8×10^6 atmospheres. This observed value of transition pressure agrees with the theoretical value.

It is worth mentioning that other elements may also exist in metallic state. For each element there is a certain transition pressure at which some electrons become free (e.g., for $H \approx 2.5 \times 10^6$, for $He \approx 90 \times 10^6$ atmospheres). The element then conducts electricity and heat well and also acquires other properties of metals. Good electric conductivity in the metallic H in Jupiter's interior explains the existence of the planet's magnetosphere.

With growing density the structure of matter and its properties become more universal. Thus for densities 50 g cm$^{-3} < \varrho < 500$ g cm^{-3} the dependence of physical properties on the atomic number Z is monotonous and smooth.

3.7.2. DEGENERATE MATTER

When density is further increased ($500 < \varrho < 10^{11}$ g cm^{-3}), all nuclei lose their electrons, so that the matter consists of free electrons and nuclear component. The main pressure of such a matter is produced by electrons and does not depend on temperature, if it is not too high. The pressure is due neither to electrostatic repulsion of the electrons, nor to their thermal motion but basically to their half-integer spin. The electrons at such high densities are degenerate as a result of Pauli principle and form a Fermi gas (Figure 3.10). The nuclei at these densities are still too far from each other (>10 Fermi) for nuclear interactions to influence properties of the degenerate matter.

A critical parameter for deciding whether a matter at a high density is degenerate or not is the degeneracy temperature T_0

$$T_0 = \frac{1}{8}\left(\frac{3}{\pi}\right)^{\frac{2}{3}} \frac{h^2}{mk} n_e^{\frac{2}{3}} = 4.3 \times 10^{-11} n_e^{\frac{2}{3}} \text{K}. \tag{3.15}$$

For very high densities the expression for the degeneracy temperature becomes more complicated. The degeneracy temperature is the temperature of transition from Fermi statistics to Boltzmann statistics. It is a function of density which is represented by a degeneracy line (degeneracy limit) in Figure 3.1. When the temperature of matter for a given density is higher than the degeneracy temperature, then its statistical properties are described by Maxwell-Boltzmann distribution. The pressure of the plasma is temperature dependent

$$p = R\frac{\varrho}{\mu} T, \tag{3.16}$$

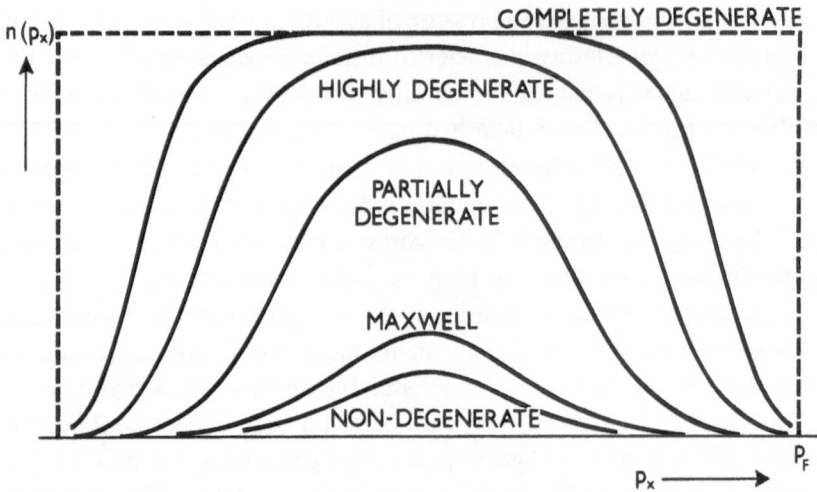

Fig. 3.10. Degeneracy. Distribution of momentum of electrons in a small volume $\Delta x\Delta y\Delta z$. Only distribution of one component of the momentum is represented (viz. of p_x). For small densities the distribution is maxwellian (ideal plasma). With the increasing number of the electrons in the $\Delta x\Delta y\Delta z$ volume the distribution approaches complete degeneracy (dashed lines) with Fermi momentum p_F as limit.

where μ is the mean particle mass in terms of the H I mass. At lower temperatures than in formula (3.15) the situation is quite different and formula (3.16) is no more valid.

Bosons with an integer spin may occupy a state in an arbitrary number. With decreasing temperature they lose their energy, until only their rest energy remains. For fermions this is not possible. According to Pauli principle a quantum level may be occupied either by one electron (fermion) or remains vacant. At zero temperature all states with momentum $p \leqslant p_F$ are occupied, while states with $p > p_F$ are vacant. The momentum p_F (Fermi momentum) depends on density of the electrons (fermions):

$$p_F = (3\pi^2)^{\frac{1}{3}}\hbar n^{\frac{1}{3}}. \tag{3.17}$$

Energy corresponding to electrons with the Fermi momentum is

$$E_F = c(m_e^2 c^2 + p_F^2)^{\frac{1}{2}}. \tag{3.18}$$

To push a new electron into the degenerate electron gas, we must endow it with an energy E_F, because only there are vacant places (cells).

By increasing density the Fermi momentum (Equation (3.17)) and Fermi energy (Equation (3.18)) are also increased, until the electrons become relativistic i.e. their kinetic energy becomes comparable to their rest energy $m_e c^2$. For densities $\varrho > 2\times10^6\,\mathrm{g\,cm^{-3}}$ most of the degenerate electrons are relativistic. The matter with $\varrho < 2\times10^6\,\mathrm{g\,cm^{-3}}$ is non-relativistically degenerate and the pressure depends only on its density:

$$p = 10^{13}\left(\frac{\varrho}{\mu_e}\right)^{\frac{5}{3}}\mathrm{dyne\,cm^{-2}} \tag{3.19}$$

where μ_e for H equals 1 and for the other elements 2. For densities $\varrho < 10^7 \, \mathrm{g \, cm}^{-3}$ the atomic nuclei remain stable.

3.7.3. Neutronization of Matter

At densities $10^7 < \varrho < 10^{11} \, \mathrm{g \, cm}^{-3}$ the cold matter consists mainly of relativistically degenerate electrons and neutron-rich nuclei. The electrons compensate for the charge of the nuclei so that macroscopically the matter is neutral.

High-energy electrons are captured by nuclei (inverse β-process). It is a weak interaction process in which a proton in the nucleus is converted into a neutron (reaction (2.31)). The process called neutronization of matter represents a further step to universality. Free protons neutronize at densities $\gtrsim 10^7 \, \mathrm{g \, cm}^{-3}$ and the very stable He nuclei at about $10^{11} \, \mathrm{g \, cm}^{-3}$ by processes

$$^4\mathrm{He} \to {}^3\mathrm{He} + \mathrm{n}; \qquad {}^3\mathrm{He} + 2\mathrm{e}^- \to 3\mathrm{n} + 2\nu_e. \tag{3.20}$$

With increasing density a large fraction of the nucleons become neutrons. The neutrons are stable. Their decay (reaction (2.19)) is not possible, because in the degenerate electron gas there are no vacant places for the decay electrons. The energy of the decay electrons is too low for the vacant places with energies above the Fermi energy (Equation (3.18)). In other words, the electrons from neutron decay have not sufficient energy to get above the Fermi threshold.

The strong interactions in the range of densities considered (10^7 to $10^{11} \, \mathrm{g \, cm}^{-3}$) are still restricted only to nuclei. The atomic nuclei at these densities are rather different from the normal nuclei we are accustomed to: the neutrons strongly prevail over protons. For example at $10^7 \, \mathrm{g \, cm}^{-3}$ the most abundant nucleus of Fe is $^{56}\mathrm{Fe}$ and by $10^{11} \, \mathrm{g \, cm}^{-3}$ it is $^{76}\mathrm{Fe}$. The neutron-rich nuclei are stable in the density range under consideration, but they would not be stable in the laboratory. However, the stability of these exotic nuclei decreases abruptly at densities $\sim 10^{11} \, \mathrm{g \, cm}^{-3}$.

At densities $10^{11} < \varrho < 10^{14} \, \mathrm{g \, cm}^{-3}$ the neutrons drip out of the nuclei and the number of neutrons outside the nuclei becomes larger than within them. The density $10^{11} \, \mathrm{g \, cm}^{-3}$ is known as the neutron drip point. At these densities the distances are so short (of the order of magnitude of one fermi) that strong interactions become a common phenomenon. All the free neutrons of the matter form one huge nucleus. Formulae traditionally used in nuclear physics are extended to this density interval to find equation of state for the neutron gas (Figure 3.11). The information on strong interactions received from nucleon–nucleon scattering experiments complement all that is known from nuclear physics about nuclear matter. The equation of state for the neutron gas $p = p(\varrho)$ relates pressure and density under the extreme conditions which are believed to exist in the interior of neutron stars. To calculate the structure of these degenerate stars (Section 4.5.3) an accurate equation of state for the neutron gas is necessary.

At still higher densities ($>10^{14} \, \mathrm{g \, cm}^{-3}$) there are three degenerate gases: relativistic electrons, non-relativistic neutrons, and non-relativistic protons. The neutrons

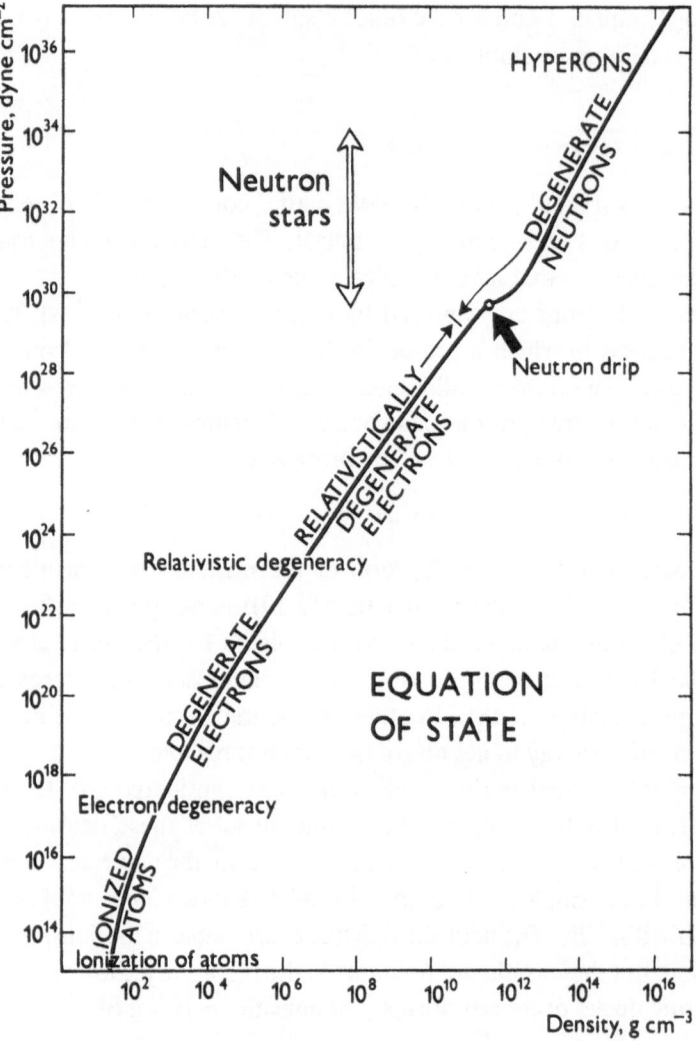

Fig. 3.11. Pressure of cold degenerate matter depends on its density as represented by the curve (called equation of state for dense degenerate matter). The equation is important for computing the structure of degenerate stars.

outnumber protons and the pressure is mainly due to the degenerate neutron gas. The density of matter in normal nuclei $(3 \times 10^{14}$ g cm$^{-3})$ is the limit in the very dense matter, above which no nuclei exist. Hyperons and mesons appear at about 10^{15} g cm^{-3}, so that the matter at these densities consist of baryons (n, p, Λ, Σ^-, Σ^+, Σ^0, Ξ^-, Ξ^0), leptons (e$^-$, μ^-) and mesons (π^-). Hyperons and neutrons become stable at these densities ($\geqslant 10^{15}$ g cm^{-3}), because their decay products would find no vacant places. Also μ^- and π^- are stable particles at these densities, because Equations (2.49) and (2.50) cannot occur due to the lack of vacant places for the produced electrons and muons.

Practically nothing is known about properties of matter at densities $\varrho > 10^{16} \text{ g cm}^{-3}$.

3.7.4. Dense matter in the universe

Gravitation is the only force which can compress ordinary matter to very high densities. Planetary interiors have relatively low pressures and densities, though sufficient to produce metallic H in giant planets. Stellar bodies contain material in a huge range of densities, from $10^{-16} \text{ g cm}^{-3}$ (as in solar corona) to $10^{15} \text{ g cm}^{-3}$ or more in central parts of neutron and hyperon stars.

It is generally accepted that the cold dense matter is a final product of stellar evolution (Section 5.4): lightweight stars end as white dwarfs, neutron stars represent final stage of middleweight stars while the final product of heavyweight star evolution is probably a black hole with matter compressed to such densities that physics today is not able to describe properties of elementary particles under such extreme conditions. It may be that even such firmly established laws as conservation of the baryon number or lepton number could lose its sense in the black hole material.

(a) *White dwarfs* were the first observed stars with high densities, in the range 10^5 to $10^{10} \text{ g cm}^{-3}$, depending on their mass. They have star-like masses and planet-like size. Therefore their densities are much higher than those of normal plasma stars, like our Sun. For example, if the entire matter of the Sun (mean density 1.4 g cm^{-3}) were compressed to the size of our Earth, its density would be about 10^6 g cm^{-3}, i.e. a mean density of a typical white dwarf. The central densities are much higher (Section 4.5.2).

There are a few thousand white dwarfs known at present. They have a very low luminosity so that only the nearest white dwarfs are known. The o_2 Eridani B is the only white dwarf for which all stellar characteristics are known with a good accuracy – viz.: mass, luminosity, color, spectrum and surface temperature. Sirius B is suspected to be a close binary itself and Procyon B is so close to its primary and so faint that its magnitude, color and spectrum are difficult to determine. However, many more binaries (about 200) are known, containing one white dwarf. This is important for the determination of mass. There are even binaries (more than a dozen) whose components both appear to be white dwarfs. Future studies of these white dwarfs with large telescopes will bring more data on the degenerate stars to improve their theory.

The frequency of white dwarfs in space gives an idea about the quantity of degenerate matter in the surrounding universe. They represent roughly 5% of all stars in the nearby universe. The degenerate matter exists in large quantities and is more abundant than the solid state.

(b) *Neutron stars* are the densest objects in the observable universe. Rotating neutron stars are generally identified with pulsars discovered in 1967. Their theory was already discussed three decades before their discovery. From observations of pulsars physical parameters of neutron stars may be deduced (energy balance of supernova remnants – Crab Nebula and Vela X remnants; the slowing down of pulsars; companions in binary systems observed as pulsating X-ray sources; red-shifted γ-ray lines from the surface of old neutron stars).

The total number of neutron stars in our Galaxy is not known. If neutron stars are formed simultaneously with supernova explosion and if the supernova frequency were 1 event per century, then the number of neutron stars in our Galaxy would be 10^8 and the neutron gas would represent a tiny fraction ($\sim 10^{-3}$) of the total galactic mass. But the 'ifs' make the estimate very uncertain.

(c) *Hot dense matter.* The dense matter in white dwarfs, neutron stars and in hypothetical hyperon stars, is at temperatures lower than degeneracy temperatures corresponding to the given density. It is called cold matter or frozen matter. If the temperature is higher than the degeneracy temperature, then it is hot dense matter. This form of matter probably existed during the first moments of our universe (Big Bang). It has often been assumed that large fractions of the hot dense matter survived in nuclei of galaxies and quasars and in exploding they liberate huge amounts of energies. The observed explosions of galactic nuclei would thus represent repetition of Big Bang in smaller scale – little big bangs.

Summary

Agglomerates of elementary particles represent matter in macroscopic sense. Its properties (state of matter) derive from properties of the constituent particles, their energy (temperature) and density. Most of the observable matter in the universe is in plasma state (stars, interstellar plasma), while neutral gas and condensate (solid and liquid states) are relatively rare.

The plasma state is changed by different processes – which may be visualized in Figure 3.1. It may be cooled to neutral gas, liquid, and solid. Plasma may be heated, e.g., by compression in stars to temperatures 10^6 to 10^{10} K to produce thermonuclear reactions or even splittering of nuclei as in last phases of thermonuclear evolution.

By a further increase of the temperature (10^{10} to 10^{11} K) nuclei are destroyed in nucleons (nucleon gas). Such an extreme state probably exists for a short time in the shock wave produced by gravitational collapse of middleweight stars, i.e., during supernova explosions.

Ions and electrons of a plasma may be accelerated from thermal velocities to a velocity approaching that of light (cosmic rays). This energetic state of matter is generated by supernovae, pulsars, radio galaxies, quasars and exploding galactic nuclei. Magnetic fields on stars, in interstellar and intergalactic space are an important factor in the acceleration processes.

By compression plasma may become cold dense degenerate matter (10^3 to 10^{11} g cm^{-3}) or very dense neutron gas (10^{11} to 10^{14} g cm^{-3}). White dwarfs are stable degenerate stars and pulsars are stable neutron stars with very fast rotation. More dense state of matter ($> 3 \times 10^{14}$ g cm^{-3}, i.e., hyperon gas or baryon gas) has been explored theoretically. At these densities different unstable particles (hyperons, mesons) become stable due to degeneracy of fermions.

The matter of our terrestrial environment has a low temperature (about 3×10^2 K or 0.03 eV) and low densities (10^{-4} to 10^1 g cm^{-3}). It consists of hundreds of thousands of different molecules. This diversity is lost when temperature and density are increased: molecules are dissociated into atoms, atoms are ionized in free electrons and ions or nuclei, nuclei are decomposed into nucleons. At high densities or temperatures matter exists in its primitive form, as free elementary particles and without any of the structural units we are accustomed to in the terrestrial environment.

CHAPTER 4

STRUCTURES

To understand things means to explain them as structures made from simpler units. Complexity of things at one level is explained by simpler entities at lower and more universal levels. Elementary particles are the most universal and simplest entities of which all things in the universe consist. Four interactions force upon elementary particles different structures. The sequence of structures ranges from elementary particles to superclusters of galaxies.

We may notice that forces important for one structure leave some small residual forces important for organization of matter at higher levels of structure. Nucleons are held together by strong interactions to form a nucleus. The electric forces are a weak residual of the nucleus – they hold electrons in orbits, dominate the atom and form a higher structure. But the electrons do not cancel electric forces of the nucleus at every place, even if they are equal to the number of protons. The residual electric forces give rise to chemical bonds which unite atoms into molecules. Residual electromagnetic forces of molecules (Van der Waals forces) organize elementary particles into higher structures: molecules to crystals, crystals to rocks, meteorites, small satellites and small asteroids. With the increasing size of structures the role of the weakest interaction – gravitation – increases and becomes dominant: large satellites, large asteroids, planets, stars, globular clusters, galaxies, clusters of galaxies.

The ensembles of particles at any structural level interact and form a system. There is a great variety in size (Figure 4.1) and in complexity. Man is much smaller but more complex than a star. The interaction (communication) between cells of an individual (through nerves and hormons) and between individual and his species (through genetic messages) is certainly very complicated; but it may be reduced to electromagnetic interactions of the same nature as those acting between ions and electrons in a star. An important feature of our universe is its unity. Everything in it consists of a few types of elementary particles interacting by four types of forces. Man is not excluded from the unity, he is a part of the universe. Stones, stars, and water are our relatives – 'brother Sun' and 'sister water' in the language of Saint Francis d'Assissi seven centuries ago.

4.1. Elementary Particles

Elementary particles are not dimensionless points. They have finite spatial dimensions and spatial structure. In a matter at very high densities ($\geqslant 10^{15}$ g cm^{-3}) the

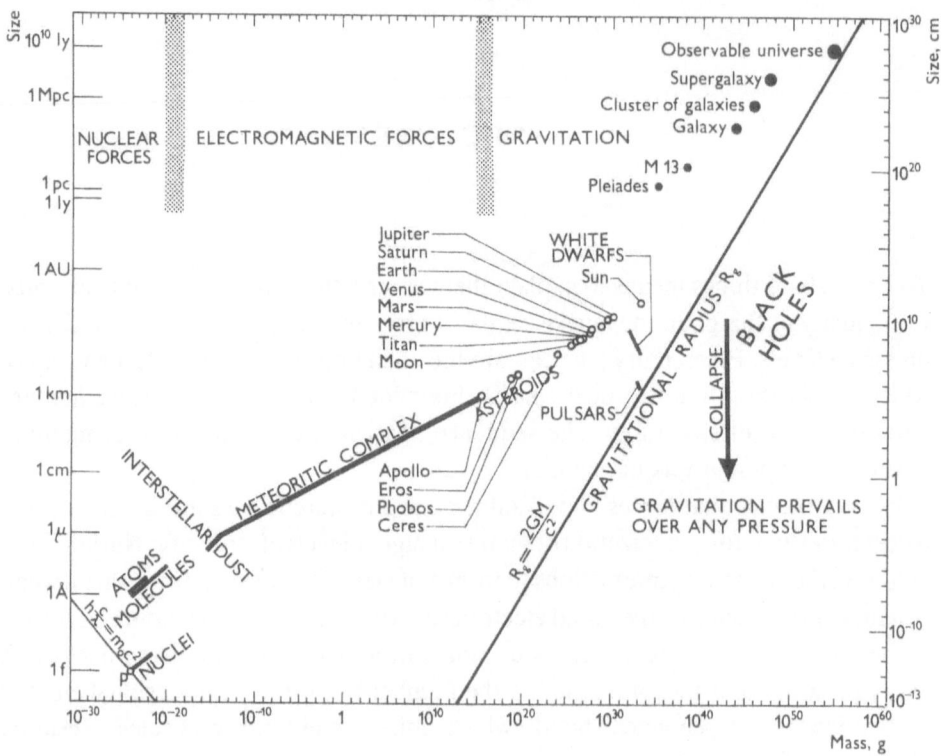

Fig. 4.1.   Mass and size of structural units in the universe.

volume per particle becomes smaller than the intrinsic volume of the particle itself; the particles cannot be hard spheres.

Physicists know the language of elementary particles as of the smallest units at all. How then to describe their internal structure? Are all the elementary particles really elementary? What does elementary nature mean? Do protons and neutrons consist of homogeneous substance or do they contain discrete and still more fundamental particles? Are varyons excited states of nucleons with energies of excitation thousand times larger than the excitation energies of nuclei? Much theoretical and experimental effort is being exerted for answering those difficult and fundamental questions.

4.1.1. QUARKS AND PARTONS

Quarks have been considered to be the sub-elementary particles with fractional charges ($-\frac{1}{3}$ e and $\frac{2}{3}$ e) and with strangeness 0 and -1 as may be seen in Table 4.1. Antiquarks $\tilde{s}$, $\tilde{d}$, $\tilde{u}$ have symmetrical charge $+\frac{1}{3}$ e and $-\frac{2}{3}$ e and strangeness 0 and $+1$. The strange name of these hypothetical sub-elementary particles comes from a play 'Finnegan's Wake' by James Joyce: "three quarks for Muster Mark".

The hadrons and reactions between them have been explained by combination of quarks and antiquarks (Table 4.1). For example, the process (1.11)

$$\pi^+ + p \rightarrow \Sigma^+ + K^+$$

in terms of quarks is

$$u\tilde{d} + uud \rightarrow uus + u\tilde{s}$$

(4.1)

which is consistent with conservation laws. The net number of a type of quarks should be conserved.

TABLE 4.1

Quarks and antiquarks are hypothetical building blocks of hadrons. They carry electric charge (fraction of the electron charge), half a unit of spin and one unit of (or none) strangeness

Electric charge Strangeness	$-\frac{1}{3}$	$+\frac{2}{3}$	$-\frac{2}{3}$	$+\frac{1}{3}$
+1				$\tilde{s}$
0	d	u	$\tilde{u}$	$\tilde{d}$
−1	s			

Examples of baryon structure:

Particle	Consists of	Total charge	Strange-ness	Spin
p	u u d	+1	0	$\frac{1}{2}$
n	d d u	0	0	$\frac{1}{2}$
Λ	u d s	0	−1	$\frac{1}{2}$
Σ^+	u u s	+1	−1	$\frac{1}{2}$
Ξ^0	s s u	0	−2	$\frac{1}{2}$
Ω^-	s s s	−1	−3	$\frac{3}{2}$

Examples of meson structure:

π^+	u $\tilde{d}$	+1	0	0
K^+	u $\tilde{s}$	+1	+1	0

Attempts have been made to discover quarks in accelerator experiments and in the universe (e.g. in cosmic rays, in solar photosphere) but in vain. Nevertheless, the idea of an inhomogeneous structure of baryons has been revived after some recent experiments in accelerators: the scattering of high energy electrons (e.g., 2×10^{10} eV electrons from the Stanford Linear Accelerator) on protons and neutrons indicate that their charge is not evenly distributed, but it is concentrated in point-like constituents called partons. They are elementary in the sense that they act like point particles with respect to electromagnetic interactions. This scattering experiment is analogical to Rutherford's discovery of nucleus with α-particles. The large deflection of α-particles passing through a gold foil indicated that the positive charge in the atom was concentrated at a point rather than diffused through the atom. Whether point-like partons are real entities and related to the hypothetical quarks remains still to be answered.

4.1.2. GEOMETRODYNAMICS AND ELEMENTARY PARTICLES

The general theory of relativity relates curvature of space and distribution of matter. What is primary – matter shaping the geometry of space or vice versa? The second eventuality is worked out in a space theory of particles, which envisages elementary particles not as foreign objects immersed into space but as structures produced out of the geometry of space. If so, the geometry of space would thus appear not as static but rather as dynamic (*Geometrodynamics* – A. Wheeler).

Particles in the theory are regarded as quantum states of excitation of the dynamic space. While the dynamics follows the deterministic evolution of the metric coefficients g_{ik} with time according to the Einstein equations, at small distances L the quantum fluctuations of the metric occur everywhere in space:

$$\delta g_{ik} = \frac{L^*}{L} \tag{4.2}$$

where

$$L^* = \left(\frac{\hbar G}{c^3}\right)^{\frac{1}{2}} = 1.6 \times 10^{-33} \text{ cm} \tag{4.3}$$

is the Planck length, the only quantity with dimension of length constructed from the constants G, $\hbar$, c which should appear in a quantum theory of gravitation yet to be constructed. The particles with size 10^{-13} cm are enormously larger than the Planck regions with size L^* and appear as 'geometrodynamics excitons'.

Whatever the nature and structure of the elementary particles may be, they themselves are fundamental structural units of the universe. In particular, the stable electron and proton as well as the quasistable neutron constitute matter at normal densities and temperatures. At very high densities neutrons and hyperons become stable constituents of matter. All particles and antiparticles – stable and unstable – played a role in the first phase of the universe evolution when all matter existed in a very hot and very dense state. The stable particles at the present time appear as fossils from the primeval hot dense matter.

4.2. Nuclei and Atoms

4.2.1. NUCLEUS

The simplest system of elementary particles is atomic nucleus (Section 2.1.1). The volume of a nucleus is proportional to the nucleon number A. For its radius R there is an approximate expression

$$R = 1.3 \, A^{\frac{1}{3}} \text{ fermi.} \tag{4.4}$$

The nuclear radii are then in the interval from 1.3 fermi to about 8 fermi for ^{238}U $(1.3 \times \sqrt[3]{238} \approx 8)$. It follows from Equation (4.4) that the density of matter in different nuclei is approximately equal and very high $(\sim 10^{14} \text{ g cm}^{-3})$.

The atomic nucleus is a tiny region of enormous density and also enormous energy of interaction. The nuclear forces do not extend beyond $\sim 3 \times 10^{-13}$ cm from the nucleons. The space of nucleus is thus closed with respect to nuclear forces, although it is open for electromagnetic and gravitational forces. Considerable changes in metric properties of space of nuclei on a scale of $\sim 10^{-13}$ cm may exist. They cannot be proved by deflection of light rays because the photons used for sounding should have an extremely short wavelength (high energy) and then they would materialize into a particle-antiparticle pair.

A nucleon is bound to its nucleus by a very large energy, about 8 MeV (Figure 2.1). When entering the community of a nucleus, the nucleon gets rid of a small fraction (mass defect Δm) of its mass ($mc^2 \approx 940$ MeV) in the form of energy $\Delta mc^2 \approx 8$ MeV. Building up nuclei from protons and liberating the binding energy Equation (2.2) is the main source of luminosity for all normal stars.

Although the binding energy of atomic nuclei is very large, their existence is not completely secured. A nucleus is splintered in smaller fragments and nucleons at very high temperatures in final stages of the evolution of massive stars (process (3.11)). Nuclei and free electrons are squeezed to neutrons by excessive compression (Section 3.7.3). When in interstellar space, a nucleus may be fragmented by high energy collisions with cosmic-ray particles. Too high temperatures and densities destroy atomic nuclei. Such destruction of nuclei occurs during gravitational collapse of middleweight stars, viz. in the collapsing stellar core and in shock waves formed by infalling envelope plasma. Nucleons can fuse to nuclei only if not exposed to extreme temperatures and densities. Still further strict lowering of temperatures and densities is required for existence of atoms.

4.2.2. ATOMS

Only a small fraction of nuclei in the universe may afford to possess an electronic envelope. The major part of matter in the observable universe is concentrated to stars which consist mainly of H and He nuclei and free electrons. The neutral atoms of H and He occur only in a thin surface layer and the heavier elements (more difficult to be ionized to bare nuclei) have low or very low abundances (Figure 4.2).

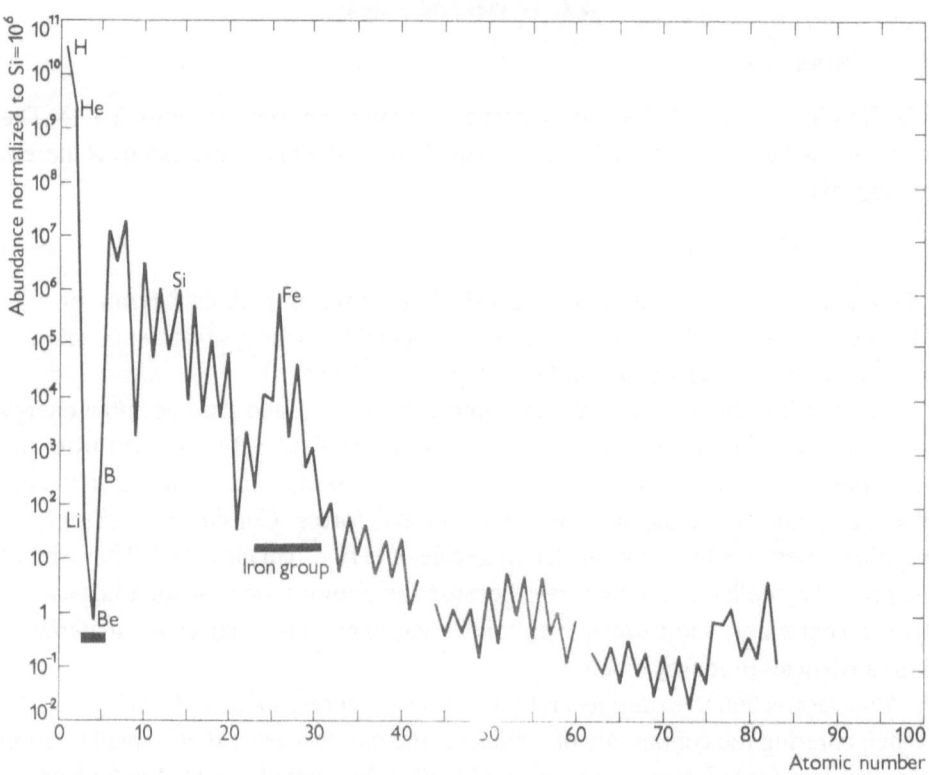

Fig. 4.2.   Chemical composition of the primitive solar nebula (according to Cameron's values). Abundancer are normalized to Si (10^6). Th and U are plotted as dots.

The binding energy of electrons in an atom is expressed in electron volts or at most in kiloelectron volts. For example the electron of a H I is bound to the nucleus with energy of 13 eV. which represents about 3×10^{-5} of the electron rest energy, while nucleons are bound with an energy of about 8 MeV. The atoms are therefore considerably less stable than nuclei. They may exist only under conditions where the energy exchange is small. In interiors of normal stars the atoms are ionized due to high temperature and in degenerate stars the electron shells are destroyed completely. In interstellar space the atoms are ionized by radiation from hot stars or by cosmic-ray particles.

The conditions in the universe are not favorable for the existence of neutral atoms. We appreciate that the temperature and pressure environment are favorable to the existence of neutral atoms on our planet and that the magnetosphere and atmosphere shield the puffed-up electron envelopes from being destroyed by high energy particles and photons. We certainly would not appreciate the reverse case, not even could we: life is based on complex molecules and they, in turn, consist of normal atoms.

4.2.3. Cosmochemistry

Atoms with the same proton number form a chemical element. In the Mendeleev periodic table the elements are ordered in a series with increasing Z and their grouping corresponds to chemical and physical properties of the elements. The huge wealth of facts of chemistry appear as a consequence of a few principles.

The study of abundances of different chemical elements (independently from the degree of ionization) is the subject of cosmochemistry. Though a modern discipline, it has already significantly contributed to our understanding of the structure and evolution of the universe.

(a) *Observations*

Cosmochemistry is based on five different methods of observation. (1) Direct chemical analysis of terrestrial rocks, of meteorites, of extraterrestrial dust (collected in ice, ocean-sediments and in atmosphere), and samples of lunar material. The same method is planned for future missions to nearby planets, comets, asteroids etc. (2) Propagation of earthquake and moonquake waves offer information about properties of deep layers. (3) Spectral analysis of electromagnetic radiation from meteors, comets, satellites, planets, Sun, stars and interstellar matter is a powerful method for studying chemical abundances of the cosmic bodies. Presence of an element in a stellar atmosphere, for example, is evident from absorption or emission lines in the spectrum of the star. Profiles of spectral lines are the main source of information on the quantity of individual chemical elements in the stellar atmosphere. There is a great variety among the million of stellar spectra photographed up-to-now mainly with objective prisma due mainly to the differences in surface temperatures of the stars. For several hundreds of thousands of stars spectral type has been determined. More than twenty thousand stars have been analyzed in detail by means of slit spectrograms. (4) Cosmic rays are registered by detectors of energy and charge of high energy nuclei at the top or outside the terrestrial atmosphere. They give information on chemical composition of cosmic rays. The information is very important, because the primary cosmic rays are the only sample of galactic matter that can be studied directly. (5) Detectors of charged particles onboard spacecraft inside and outside our magnetosphere give information on chemical composition of ionosphere, radiation belts, of the quiet and disturbed solar wind.

(b) *Abundances of Elements of the Solar System*

There is a marked similarity in the chemical composition of the terrestrial, lunar, meteoritic materials and of extraterrestrial dust. The similarity of elemental abundances supports the idea that the Earth, the Moon, meteorites and interplanetary dust in general have a common origin. They were formed from a primitive nebula with other bodies of the planetary system (Section 5.5).

The original chemical composition of the nebula has been conserved in the solar atmosphere. It is similar to that of the Earth, the Moon and meteorites except for abundance of H and He. Our Earth, other Earth-like planets and smaller bodies of the planetary system did not possess a sufficient mass (i.e. gravitation) to retain volatile elements such as H and He. These bodies represent a non-volatile fraction of the primitive nebula.

On the massive giant planets like Jupiter and Saturn, the H and He are very abundant. The giant planets have a chemical composition similar to that of the Sun, because they did not lose the light volatile elements.

The Sun has a chemical composition which corresponds best to the primitive nebula. Its atmosphere consists of 98% of H and He, and only 2% of all the other elements. Figure 4.2 represents elemental abundances in the Sun, Earth and meteorites (the low abundance of light gases on Earth and in meteorites is disregarded). It is a composite abundance curve which represents chemical composition of the primitive nebula.

The major constituents of Jupiter and Saturn are H and He, of Uranus and Neptune also C, N, O while the terrestrial planets consist mainly of Mg, Si, Fe. With increasing distance from the Sun, there is an overall tendency for an increase of the volatile (i.e., easily evaporating) elements with respect to the abundance of the non-volatile elements.

The atoms of our solar system are more than 5.5×10^9 yr old, older than the Sun. Only slight changes in their abundances have been caused by decay of radioactive elements and by cosmic rays. The fusion of protons into α-particles (^{4}He) in the solar interior is a more significant chemical change, but due to lack of a deep reaching convection it does not influence the elemental abundances of the solar atmosphere. Hence the curve of the solar-system abundances (Figure 4.2) 'remembers' the elemental composition of the primitive nebula from which the solar system developed. It is a galactic heritage in the sense that the atoms of the nebula had been produced by stars of presolar generations from primeval H.

It is generally believed that also for most other stars their atmospheric composition reflects the original chemical composition of the interstellar matter out of which the stars were formed.

(c) *Abundance Curve*

Most stars have composition similar to that of the abundance curve for the solar system. The curve in Figure 4.2 approximately represents a cosmic elemental composition. It represents a norm with which other abundances are compared and considered normal or anomalous. The curve has several important features:

(1) A gradual decrease towards higher atomic numbers. Roughly speaking, the heavier elements are less abundant then the light elements.

(2) Nuclei with even atomic numbers are more abundant than neighboring nuclei with odd atomic numbers. Particularly abundant are the nuclei which are composed of α-particles (except ^{8}Be).

(3) The most abundant element in the Sun and in the whole universe is H. It is the oldest element in the universe, because its provenance dates back to Big Bang. Its atoms represent the building units for all the other elements.

(4) The next abundant element is He. It has been produced by H burning in stars and a certain part (the magnitude of which is still disputed) may have originated much earlier, i.e., in Big Bang. The ^{4}He nucleus (α-particle) consists of two protons and two neutrons, the nuclear forces are saturated (as evidenced by zero spin of the α-particle) and hence the nucleus is extraordinarily stable. There is a less stable He isotope ^{3}He which is about 10^4 times less abundant than ^{4}He.

(5) The elements Li, Be, and B have very low abundances. Their binding energy is relatively low (Figure 2.1). The binding energy per nucleon in ^{8}Be is 7.06 MeV and in ^{4}He 7.07 MeV, so that by splitting Be into two α-particles one obtains energy. ^{8}Be is an unstable nucleus with half-time 10^{-16} s. The three light elements have been formed either in Big Bang (as Li) or by spallation of heavier nuclei in collisions with cosmic rays. In stellar interior they are fast burnt to He.

(6) The abundances of elements in the neighborhood of Fe are two or three orders of magnitude higher than if the curve were smooth. The peak is due to the fact that the nuclei have the highest binding energy (Figure 2.1). They are most stable and have the best chance to survive the final stages of stellar evolution when their neighbors in the periodic table will have been destroyed by very high temperatures.

(d) *Interstellar Abundances*

The spectrum of some stars (e.g., of the star ζ Oph) contains interstellar lines of Li, Be, Na, Al, K, Ca and Ti. The UV spectrum of the star ζ Oph (observed by the Copernicus satellite) indicates the presence of H, B, C, N, O, Mg, Si, P, S, Cl, Ar, Mn and Fe in the interstellar space between the star and us. The observed abundances of the elements with a high condensation temperature (e.g., Ca, Ti, Al, Fe) are very low. This suggests that the heavy elements have condensed into interstellar grains. On the other hand H and He cannot condense under interstellar conditions.

(e) *Cosmic Ray Composition*

Abundances of nuclei in galactic cosmic rays are remarkably similar to the abundances in the solar system. Both are compared in Figure 4.3. The differences consist in:

(1) The nuclei Li, Be, B, F and from Cl (atomic number 17) to Mn (atomic number 25) are abundant in cosmic rays though rare in the solar system. They are produced by spallation reactions during transport through interstellar medium. (Figure 4.4 represents a similar process in emulsion.) Their abundances may be used for determination of the mean age of cosmic rays.

(2) Hydrogen and He are considerably depleted, being one order of magnitude less abundant in cosmic rays than in the solar atmosphere.

(3) The large abundances of heavy nuclei, occurrence of actinide nuclei (elements with atomic numbers $Z \geq 90$) and of transuranium elements indicate that production of heavy elements and their acceleration to relativistic energies is probably due to the same process. Supernovae seem to do both: their explosion produces intense neutron fluxes for building heavy nuclei by r-process (explosive nucleosynthesis) and it

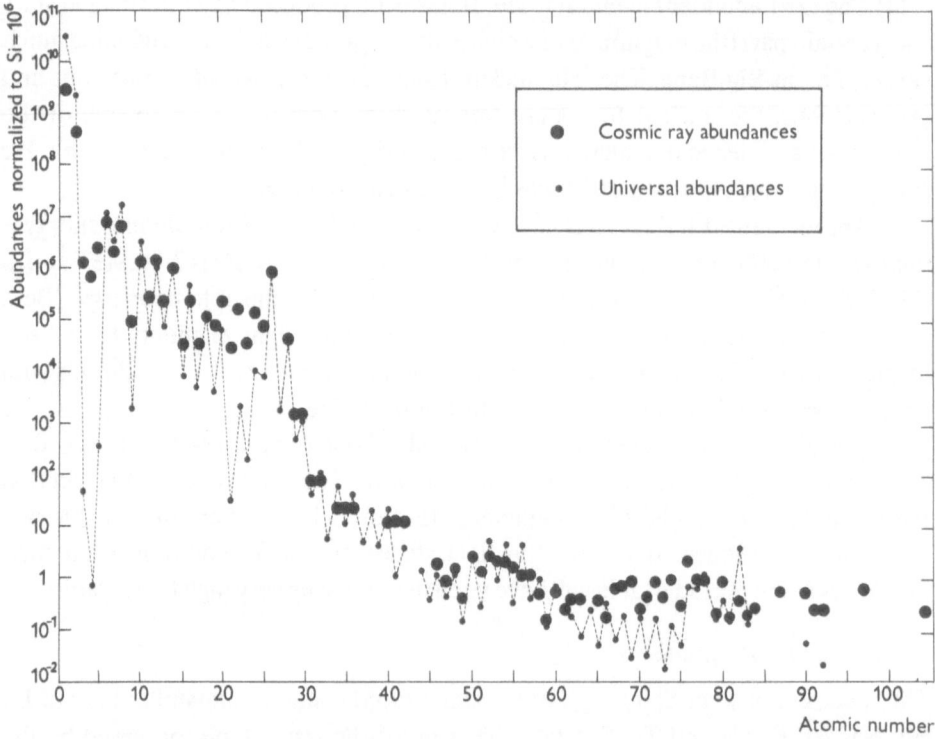

Fig. 4.3. Cosmic ray abundances normalized to Si (10^6) (according to Price). For comparison the solar system abundances (see Fig. 4.2) are plotted. Very high abundances of Li, Be and B in cosmic rays may be noticed.

accelerates them to cosmic-ray energies. Explosions of central regions of galaxies are much more violent processes. Radio galaxies are remnants of such immense explosions; their synchrotron radiation indicates that cosmic rays are produced in the explosions of galactic nuclei.

(f) *Chemical Aging of the Universe*

The nucleosynthesis in supernova explosions and in exploding galactic nuclei is one of several ways in which nuclei are built up from protons and neutrons. The conditions in Big Bang were favorable for synthesis of He and some other light nuclei (the process is known as cosmological nucleosynthesis). Non-violent evolution of stars contributes substantially to increase the abundances of many elements. These and some other processes explain the major abundance features in Figure 4.2.

The chemical evolution of matter in the universe is at least qualitatively understood. The occupancy of the Mendeleev periodic table after Big Bang was unusually simple: all matter was represented by H and (maybe with some) He. The processes of nucleogenesis have built up heavier nuclei until the present abundances (represented approximately by Figure 4.2) have been reached. But the process of chemical aging

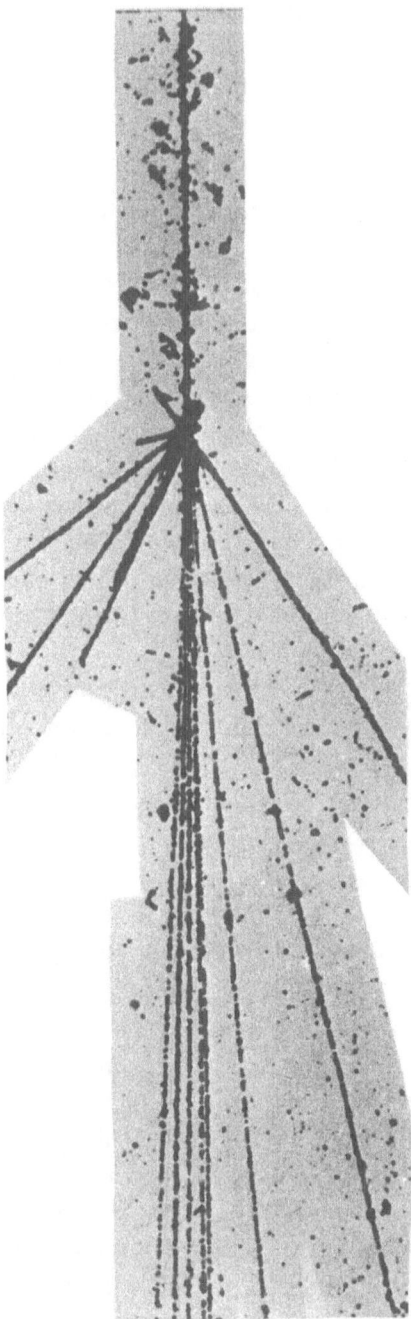

Fig. 4.4.   Fragmentation of Mg nucleus into six α-particles, as a result of its collision with a nucleus in the emulsion. The primary Mg nucleus arrives from above (a heavy track) and the six particles continue nearly in the same direction. The other tracks are fragments of the emulsion nucleus (C. F. Powell). Similar processes occur in the interstellar space – producing light nuclei in cosmic rays, among others Li, Be, B.

continues, i.e. the cosmic material is shifting from the left upper corner downwards and to the right part of the periodic table.

Atoms built up by different nucleosynthetic processes combine into molecules where temperature and pressure conditions are favorable.

4.3. Molecules and Dust Particles

Molecules are the smallest association of atoms that may be regarded as chemical units of structure. They are held together by electromagnetic forces. The number of atoms in a molecule is from two (e.g. H_2, CO, etc) to millions of atoms in some mammoth organic molecules exceptionally found in nature.

There is an essential difference between structure of molecules and that of atoms. In atoms the electrons move in the electric field of one nucleus only, whereas in a molecule they can move in the field of several nuclei. The motion of an atomic nucleus means translational motion of the whole atom. In a molecule, in addition, the nuclei can move in respect to each other. These motions are quantized and transitions between the different states of vibration or rotation give rise to complex molecular spectra.

In the universe molecules occur in relatively cool places with low pressure, due to their low binding energy which is of the order of electron volts. Molecules are found on planets, meteorites, comets, satellites, sunspots, interstellar matter and atmospheres of cold stars. The conditions on our Earth are without any doubt a paradise for the chemists.

4.3.1. MOLECULES ON PLANETS AND ON THE MOON

(a) The Earth

The lithosphere, hydrosphere, biosphere and atmosphere of our Earth consist essentially of molecules. The lithosphere consists mainly of SiO_2 (55%), Al_2O_3 (15%), Fe_2O_3 and FeO (8%), CaO (9%), etc. The hydrosphere represents about 10^{24} g of water molecules. The atmosphere consists of two major constituents, viz., N_2, O_2 and of several lesser ones: CO_2, Ar, Ne, He, CH_4, N_2O, H_2, O_3, Kr, Xe. Its total mass is estimated to be 5×10^{21} g. The biosphere represents about 10^{19} g. The total organic compounds produced by the biosphere represent about 1.5×10^{17} g yr^{-1}. Roughly speaking, the biosphere consists mainly of water (about 50% for the woods and 99% for some marine invertebrates).

(b) The Moon

The samples of the lunar surface material contain inorganic molecules known on the Earth, such as SiO_2, FeO, CaO, TiO_2, Al_2O_3, MgO, Cr_2O_3, Na_2O, K_2O, P_2O_5. Their abundances vary from sample to sample. Minerals such as pyroxene, orthoclase, ilmenite, albite, olivine, anorthite, diopside, hypersthene, apatite and feldspar belong to the main constituents of massive rocks as well as of lunar soil. The sedimentary and metamorphic rocks are absent, due to absence of liquid water. Some organic molecules have been identified in the lunar material. The samples from the landing site of Apollo 16 contain molecules HCN, CO, CH_4, H_2O, H_2S and SO_2. They may have been brought there by a recent impact of a comet.

(c) *The Atmosphere of Venus*

The atmosphere of the planet Venus consists of molecules CO_2 ($\approx 97\%$), O_2 ($\leq 0.1\%$), N_2 ($\leq 2\%$) and H_2O (6 to 11 mg l^{-1}). These results of Venera spacecraft 5 and 6 show that the Venus atmosphere has a quite different composition from that of the terrestrial atmosphere. The oxygen is probably a result of photodissociation of CO_2 and H_2O. Its low concentration indicates that it strongly recombines with rocks of the planet's surface. The amount of water seems to be large enough to imply that the clouds of Venus are composed of condensed water. Other suggestions (e.g. droplets of H_2SO_4) have been forwarded for the Venus clouds. The composition of the surface material of Venus is at present not known (incomplete analysis has been done by Venera 8).

(d) *The Atmosphere of Mars*

The CO_2 is also the main constituent of the Martian atmosphere ($\geq 70\%$). There are observational indications that polar caps of our outer neighbor probably consist of solid CO_2 under which is a layer of frozen H_2O. There is spectroscopical evidence for water vapor in the Mars atmosphere; its amount depends on position and time. Some water may be trapped as permafrost. Liquid water may exist in low-lying areas where the temperature is favorable. As for the surface composition, it is hoped that in the near future entry probes (e.g. Viking) sent to Mars will provide more precise information.

(e) *The Atmospheres of Giant Planets*

The atmosphere of Jupiter consists of molecules H_2, CH_4, NH_3, HD, CH_3D, C_2H_2, C_2H_6, H_2O and of He atoms. The CH_4 and NH_3 molecules exhibit strong absorption bands in the IR, so that their concentration in the planet's atmospheres could be determined. The H_2 has lines at 8200 Å and 6368 Å the strength of which gives an estimate for the amount of H_2 on Jupiter.

The upper layer of clouds of the giant planets should be a composite mixture of NH_3 and NH_4SH. The color changes in the Jupiter atmosphere indicate that certain molecules are formed at one place and then transported to other places where they decay.

The Jovian atmosphere has been simulated in laboratories by cooling a mixture of H_2, CH_4 and NH_3. A weak electric discharge in such an artificial atmosphere produced a series of organic molecules, such as acetylene, ethylene, etan, and methyl-cyanide. Some of the products are colored. Whether the laboratory results are consistent with the real atmosphere of Jupiter (and other giant planets) will remain a question until the entry probes descend into its atmosphere and perform an analysis *in situ*.

4.3.2. MOLECULES IN METEORITES AND COMETS

A sample of cosmic matter in a chemist's laboratory is the most reliable source of information for cosmochemistry. Meteorites in particular have been carefully analyzed.

(a) *Anorganic Molecules in Meteorites*

Simple anorganic molecules such as SiO_2, SiC, Fe_3C, FeS, MgO, FeO, Al_2O_3, CaO, TiO_2, MnO, K_2O, Na_2O, CoO, V_2O_5, P_2O_5, and Cr_2O_3 have been found in meteorites. Many minerals found in meteorites occur also in terrestrial rocks (e.g., serpentine, olivine, magnesite, ilmenite, apatite etc.). But there are some minerals that occur only in meteorites, such as osbornite (TiN), oldhamite (CaS), daubreelite ($FeCr_2S_4$), lawrencite ($FeCl_2$), merrillite ($Na_2Ca_3(PO_4)_2O$) and a few others which have not been found in terrestrial lithosphere.

(b) *Carbonaceous Chondrites*

Special attention has been paid to stony meteorites with small spherules called chondrules (from the Greek '*chondros*' – grain). Their size ranges from a pinhead to a pea. Chondrules are embedded in a mass called matrix. The stony meteorites with chondrules are called chondrites. The Swedish chemist Berzelius already discovered in 1834 that some chondrites contain organic molecules (hydrocarbons). The chondrites rich in hydrocarbons are called carbonaceous chondrites.

About twenty carbonaceous chondrites are known. The largest of them, weighing 16 kg, is the size of a human head. The best known carbonaceous chondrites – called after the place of their fall – are Orgueil meteorite, Murchison meteorite and Murray meteorite.

What makes the study of carbonaceous chondrites especially attractive is their content of many complex organic molecules, in particular of aminoacids. More than twenty different aminoacids have been recently identified in the Murray and Murchison meteorites. They are a rare variety of carbonaceous chondrites (type II) containing as much as 3% of C. Some of the amino acids identified are not found in terrestrial proteins (e.g., sarcosine CH_3NHCH_2COOH, β-alanine $NH_2CH_2CH_2COOH$). Several of the amino acids have almost equal abundances of their D and L isomers; the isomers differ in arrangement of atoms in their molecules, so that the polarized light passing through them is rotated in the right hand (D) or the left hand (L) sense. These and some other properties indicate that the amino acids in carbonaceous chondrites are not terrestrial contamination, but that they are indigenous to the meteorites: they are thus of extraterrestrial origin.

Amino acids are synthesized in living organisms to proteins which are of basic importance for life. A vivid discussion continued for some time about the origin of the complex molecules found in carbonaceous meteorites: are they from living matter or not? Even if they have been formed by non-living matter, their presence in meteorites helps to understand the history of meteorites and of the whole solar system. They would prove that complex chemical synthesis can take place in an extraterrestrial environment (Section 5.6). By irradiation of a mixture of H_2O with CO_2, CO, CH_4, SO_2 and some other molecules which exist in volcanic gases, complex organic molecules may be produced – similar to those identified in carbonaceous meteorites.

The carbonaceous chondrites are supposed to be of cometary origin.

(c) *Molecules in Comets*

Groups of atoms (radicals and ions) observed in comets (C_2, CH, CN, NH, NH_2, OH, CH^+, CO^+, CO_2^+, N_2^+, OH^+, C_3) are molecular fragments that can exist only in medium without collisions. They are apparently formed by photodissociation of (as yet unobserved) stable parent molecules, which are steadily produced from the cometary nucleus.

The cometary nucleus is probably a snow ball or icy conglomerate of the parent molecules: H_2O (parent of OH and OH^+), CO (of CO^+), NH_3 (of NH and NH_2) and of molecules observed in interstellar space such as $CH_2{=}O$ (parent of CH and CH^+), $CH{\equiv}C{-}CH_3$ (of C_2 and C_3), $H{-}C{\equiv}N$ (of CN) and some others. The good correspondence to all cometary radicals supports the idea that comets are of insterstellar origin (Section 4.4.3).

4.3.3. MOLECULES IN SUN AND IN STARS

(a) *Solar Molecules*

Some stable simple molecules are found in cool spot umbrae, with temperatures $\leqslant 4000$ K. For example: OH, NH, O_2, CH, CN, CO, C_2, MgH, SiH, TiO, CaH, BH, ScO, BO, AlO, ZrO, YO, SiF, H_2, NO, SiO, and N_2. In the warmer photosphere only the most stable molecules (such as CH) can survive.

(b) *Stellar Molecules*

Lines and bonds of molecules with a large dissociation potential are found in spectra of all stars except in O, B and A stars. The molecules are relatively abundant especially in atmospheres of low-temperature stars (~ 3000 K), so that their spectral bands and lines are easy to observe. Among the molecules which best resist dissociation one finds compounds of H, C, and O. Thus for example:

carbon stars have C_2, CN, CH, NH, SiC_2, C_3, H_2O;

S-type stars contain ZrO, LaO, YO, SiH, TiO and

M stars have in their atmospheres molecules TiO, CN, CH, VO, MgH, SiH, AlH, ZrO, ScO, YO, CrO, AlO, BO, SiF, SiN, C_2, (and CaOH in M-dwarfs).

The amount of molecules in some stars is variable – especially in variable stars. For example the cool variable giant χ Cygni changes its brightness with a period of 409 days. When it is faintest, the AlH bands are very intense, as a result of massive recombination of Al and H. We may thus observe chemical processes on distant stars.

4.3.4. INTERSTELLAR MOLECULES

In optical spectra of some stars lines of interstellar molecules CH, CN and CH^+ have been identified. The lines appear in absorption and are very sharp, in contrast to lines of the background star which are always broadened.

Molecules H_2 have been identified in the UV spectrum of stars (Figure 4.5). Carbon monoxide with large dissociation energy (11 eV) being the most stable diatomic molecule is expected to be the most abundant molecule in interstellar space after H. It has been detected in the UV spectra of some stars by spectrographs outside the Earth's atmosphere.

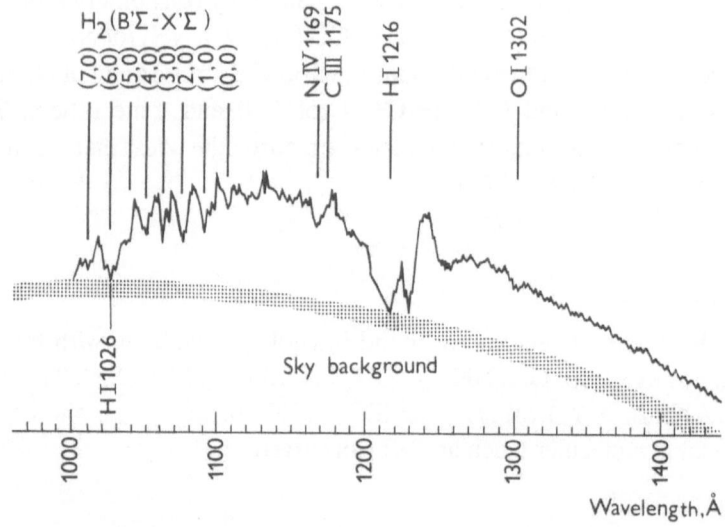

Fig. 4.5.   Ultraviolet spectrum of ξ Persei with absorptions of interstellar H and H_2 (Carruthers).

Radio observations have recently yielded very exciting results. Radio astronomers discovered more than thirty interstellar molecules containing atoms and isotopes of H, C, O, N, S and Si (Figure 4.6). Most interstellar molecules are observed in emission, some in absorption and a few in both emission and absorption. Some of the interstellar molecules (e.g., OH, CN, CH_2S) are unstable and transient under laboratory conditions. Some polyatomic molecules in interstellar space have as many as seven atoms (CH_3CHO, CH_3C_2H).

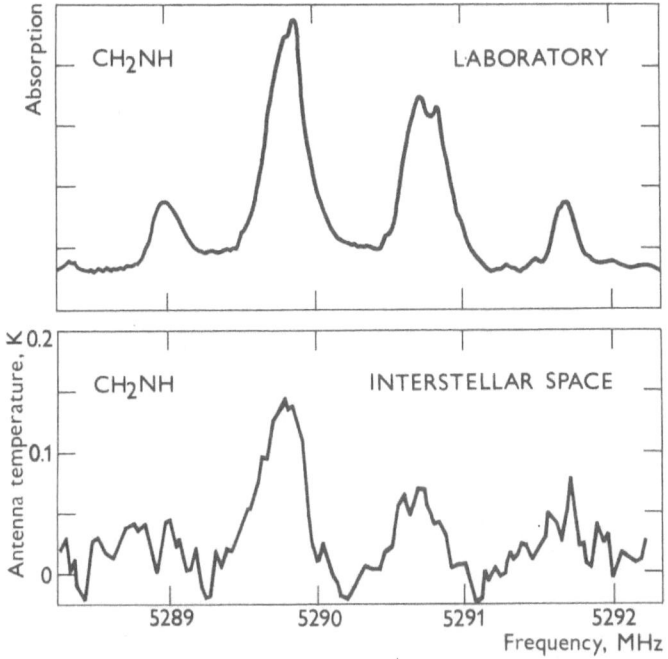

Fig. 4.6. Identification of interstellar molecules (CH₂NH) by comparing their laboratory spectrum (Monash University) with radio spectrum of an interstellar source in Sagittarius (Parkes radio telescope).

The interstellar molecules are associated with H II regions, with IR objects and dense cool clouds. The clouds are composed mainly of molecules and dust grains. The dust grains seem to be important for formation of some molecules – as in the case of H_2:

$$H + H + grain \rightarrow \text{excited } H_2 + grain \rightarrow H_2 + \text{heated grain} \qquad (4.5)$$

At the same time grains shield molecules from photodisruption; in normal interstellar space the H_2 molecules can survive only about 10^3 yr. The dense molecular clouds are presumably sites of star formation. The fact that organic molecules are abundant in dense clouds – precursors of formation of stars and planetary systems – has deep biological implications.

The prebiotic molecules, such as formamide, formic acid, methanimine, cyanoacetylene) might have served as material for biological evolution, after having been transported from the harsh interstellar medium (Figure 4.7) into a planetary atmosphere. Thus for example the formic acid (HCOOH) and methanimine (H_2CNH) may produce a simple amino-acid, viz., glycine NH_2CH_2COOH. From a mixture of abundant interstellar molecules (such as H_2O, HCN and H_2CO, supposed to be constituents of primordial planetary atmospheres) complex organic compounds have been synthetized in laboratories: alanine, aspartic acid, adenine, glycine etc. Were these organic molecules already synthesized in dense clouds before their

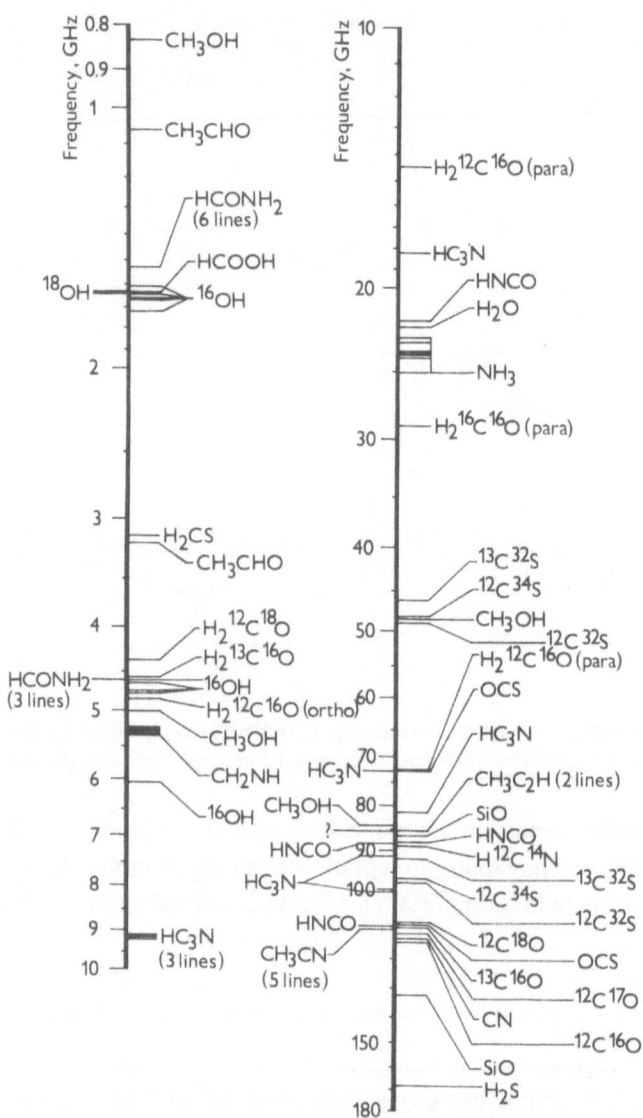

Fig. 4.7. Interstellar molecules and their radio lines.

condensation into stars and planets? The question may be answered by future observations.

The chemistry of interstellar molecules has little in common with the chemistry on the Earth's surface. Here, the hierarchy of structures – molecules – macromolecules – megamolecules – subcellular particles – virus – cell – multicellular organism can exist thanks to the energy from the Sun and shielded by the magnetosphere and atmosphere against destruction. The whole terrestrial biosphere represents only an extremely tiny fraction of the molecules in the universe, but one prominent in complexity. In the biological hierarchy the electromagnetic interactions have the same basic importance as in any other molecule in the universe.

4.3.5. INTERSTELLAR GRAINS

Larger agglomerations of elementary particles than molecules range from tiny dust grains to clusters of galaxies. Their mass ranges from $\sim 10^{-22}$ to $\sim 10^{47}$ g which means from $\sim 10^2$ to $\sim 10^{71}$ nucleons. The bodies at the light-weight end of this huge interval of masses are held together by electromagnetic interactions, not different from those encountered in molecules. They are in solid state, with different degrees of ordering in crystals and amorphous state.

The smallest bodies in the universe – next to molecules – are among interstellar particles. The name 'dust grains' is used for them, though 'smoke particles' would be more appropriate. They are probably formed like smoke, from the gas on condensation nuclei and not like dust by breaking up larger bodies. The idea of a genetic connection between interstellar dust and gas is supported by a common distribution of both in interstellar space.

The formation of dust particles from interstellar gas is supposed to proceed in three steps: (1) Formation of simple molecules (CH, CN, SiO_3, . . .); (2) Growth from the molecules to aggregates of ~ 10 or more molecules. These molecular aggregates are condensation nuclei for future grains; and (3) Growth of condensation nuclei to grains. Another possibility for step (2) would be that the condensation nuclei are formed in cool giant stars and leave the stellar atmosphere like smoke particles leave a flame. Thus the atmosphere of carbon stars may be a proper site for graphite grain formation. They are irregular variables with spectra dominated by CH, CN, and C_2 molecules. The temperatures at minimum of the stars (~ 1500 K) are low enough for graphite flakes formation. The condensing flakes are pushed away from the parent star by its radiation pressure.

There are about 10^5 carbon stars in our Galaxy and they may produce the observed interstellar grain density ($\sim 10^{-26}$ to 10^{-27} g cm^{-3}) in about 10^9 yr. The interstellar medium is drained of its dust content by formation of stars. The characteristic time of this destruction process, however, is also $\sim 10^9$ yr: the dust density thus remains in steady state. The oxygen-rich Mira-Ceti stars (with molecules SiO_2, SiO, CO_2, CO, . . .) and supernova explosions have been also considered as possible sources of solid grains.

Though much observational data have been accumulated about extinction of starlight, its polarization, its scattering and on IR emission by grains, very little is still known about the size, shape and chemical composition of the interstellar dust grains. The observed extinction and scattering properties indicate that the particles may be composed of refractory materials – such as graphite, iron, quartz and silicates. The polarization of star light is explained by an elongated shape of grains which are oriented by some mechanism (e.g., by interaction with interstellar magnetic field, with soft X-rays, with ambient gas).

The shape of tiny interstellar particles may be quite irregular (like that of smoke particles), and they may lack a compact crystal structure. The extinction depends both on the size of particles ($\sim\lambda^{-1}$ extinction law) and their refractive index; determination of size from the extinction curve thus becomes rather uncertain.

There are still few firmly established observational facts on interstellar dust grains, too few for a clear description of their geometrical, chemical and physical properties. Much better is the situation with the solid particles in our planetary system. The sources of information are more numerous and more reliable than in the case of interstellar particles.

4.4. Bodies of Our Planetary System

In the gravitational field of the Sun many bodies move, ranging in size from tiny grains up to the giant planet Jupiter. They are classified as meteoritic complex, asteroids, comets, satellites and planets. All together they form our planetary system. If the central body – the Sun – is included, then the whole system is called solar system.

4.4.1. METEORITIC COMPLEX

The lightest members of the planetary system range in size from tiny grains consisting of a few molecules ($\sim10^{-7}$ cm) up to the smallest asteroids (a fraction of a kilometer in diameter). Their ensemble is called meteoritic complex or solid component of interplanetary matter. The particles themselves are sometimes called meteorites, other times, meteoroids. The term meteoroid has been proposed to differentiate the interplanetary bodies from the minerals of extraterrestrial origin, i.e., meteoroids fallen on the Earth. If one term (meteorite) is used for both meanings, the danger of confusion is minimized by the context.

(a) Meteoric Dust

The smallest particles of the complex ($\leqslant10^{-7}$ g) survive entry into our atmosphere. There they float with the fine dust particles left over from evaporation and ablation of larger particles. The meteoric dust of either origin falls slowly towards the Earth's surface. There it is collected on collecting plates carried by balloons, at isolated places, from old arctic snow and ancient ice deposits in Greenland, from ice cores at the South Pole, from sedimentary deposits and ocean-bottom sediments. The total amount of the meteoric dust (floating and drifting continuously downwards) in the atmosphere is about 10^6 t. The daily mass accreted by the Earth from the meteoritic complex is roughly 400 t.

(b) Meteorites

If the mass of a meteoroid is large enough, it is not completely evaporated and ablated during its passage through the terrestrial atmosphere. The remaining piece impacting on the Earth's surface is a meteorite. The mass of the meteorite represents only a certain fraction of the original meteoroid mass entering the atmosphere. The losses are very sensitive to velocity. Thus for 12 km s^{-1} about 0.9 of the original mass impacts on ground, while for 20 km s^{-1} the remaining meteorite represents only about 1% of the original mass.

The bulk of the meteorite material is either Ni–Fe alloy (mean density about 8 g cm^{-3}) or silicates (3 to 5 g cm^{-3}) with inclusions of Ni–Fe and small chondrules. The impact of meteorites upon the solid surface forms impact craters, be it on the Earth, other planets, and satellites (Mars, Mercury, Moon, Phobos). Their size depends, among others, on the mass and velocity of the meteorite. The craters permit an estimate of the impact rate and size distribution of the meteoritic complex in the past.

Meteorites represent only a small fraction of the total material of the meteoritic complex acquired by the Earth, i.e. about 10^{-4} to 10^{-3}. To the present time more than two thousand meteorites have been found. Most of the material falling on the Earth from the complex is in the form of fine particles.

(c) Meteors

Particles larger than meteoritic dust (roughly $>10^{-7}$ g) and smaller than meteoroids producing meteorites (limit strongly dependent on entrance velocity) – are destroyed when entering the Earth's atmosphere. Their kinetic energy is transformed into fusion, ablation, light and ionization, producing meteor phenomena at heights of 80 to 120 km.

From a certain size (more than about 0.001 g for 40 km s^{-1}) the light emission may be seen by our eye as luminous phenomenon (optical meteor). The brightness of the meteor depends on the mass and velocity of the infalling particle (from 10 km s^{-1} to 70 km s^{-1}). The particles with smaller masses have not sufficient energy to be seen by a naked eye. They are observed by a telescope (telescopic meteors) and by radars on meter wavelength (radar meteors).

The meteor photography gives information on hourly flux, size distribution, velocities and orbits and also on chemical composition if spectra are photographed. The data relate mainly to the middle-size particles of the meteoritic complex, the small sizes are studied by other methods (radar, zodiacal light, detection by spacecraft – Figures 4.8 and 4.9). The population of larger particles in the meteoritic complex is studied by the effects which impacts of meteorites or small asteroids produced – in the form of impact craters – on the surfaces of the Moon, Mercury or Mars.

(d) Meteoritic Complex

Figure 4.10 gives an idea of how the meteoritic complex is populated. It is an approximate representation of observational data. The real distribution will depend on the position in interplanetary space and will probably change with time; there are for example streamers of meteoroids left over by decaying comets.

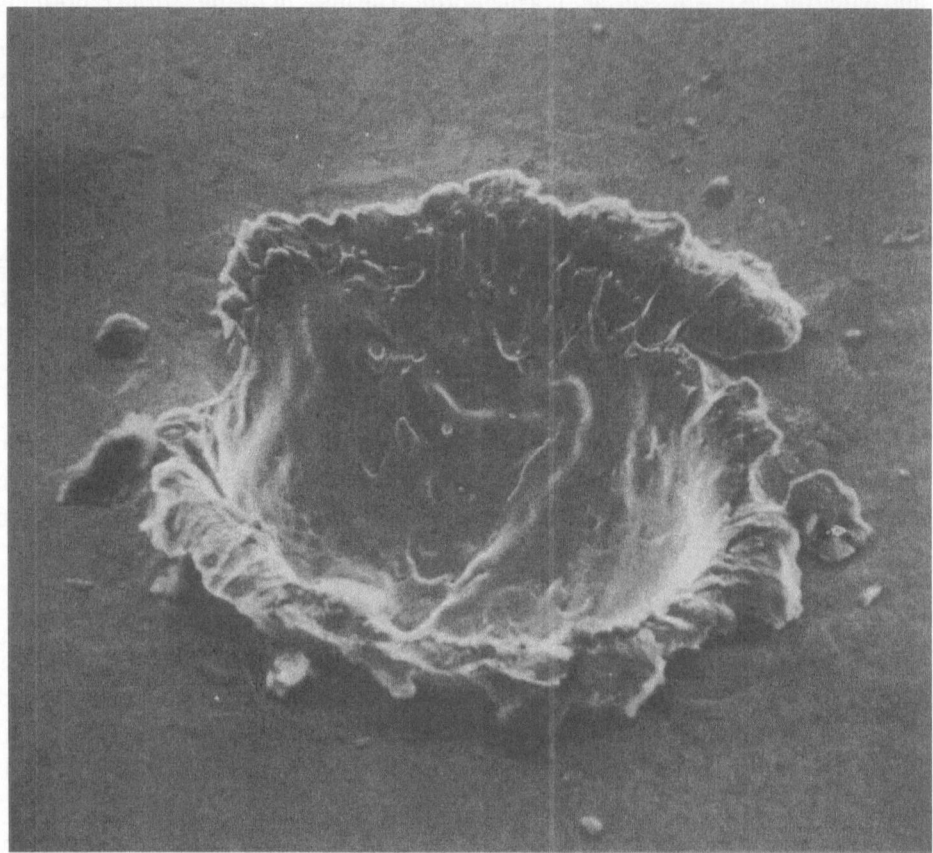

Fig. 4.8. A small crater in polished stainless steel, produced by impact of a tiny interplanetary dust particle. Width of the crater is 160 μm and its depth 65 μm (The Dudley Observatory Experiment S-10, Gemini Program).

The meteoritic complex is not a dead system. Different destructive processes (such as sublimation near the Sun, ejection by encounters with large planets, erosion, collisions, light pressure eliminating the smallest particles <1 μm) must be compensated by sources of meteoritic complex (comets, asteroids) to maintain its quasi-equilibrium (Section 5.5.4 and Figure 4.10).

The total mass of the meteoritic complex is estimated to be 2.5×10^{19} g. It is completely replenished in about 1.7×10^5 yr. About 10 t of new material is fed each second into the complex. Comets and asteroids are the main sources but their relative role is yet to be determined. Large meteoroids ranging in mass from a few kilograms to about 10^6 kg are probably of asteroidal origin formed by breaking-up of the Apollo-type asteroids (i.e., asteroids with perihelia inside the Earth's orbit.

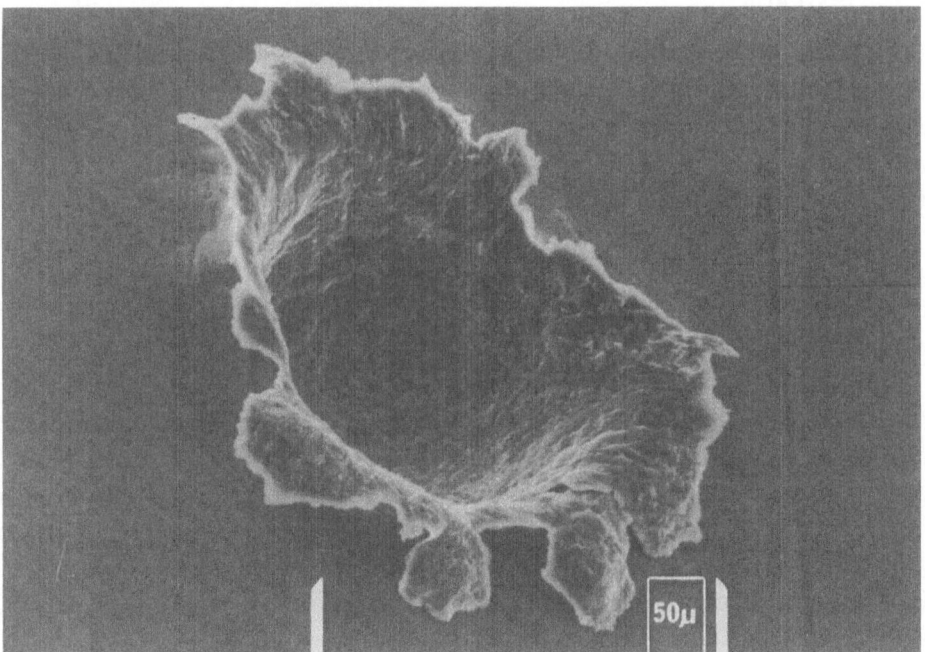

Fig. 4.9. Micrograph of a tiny impact crater in Cu (The Dudley Observatory Experiment S-149, Skylab). The size of the incident particle was comparable to the size of the impact crater. It is evident that the metal was melted by heat derived from the kinetic energy of the impact.

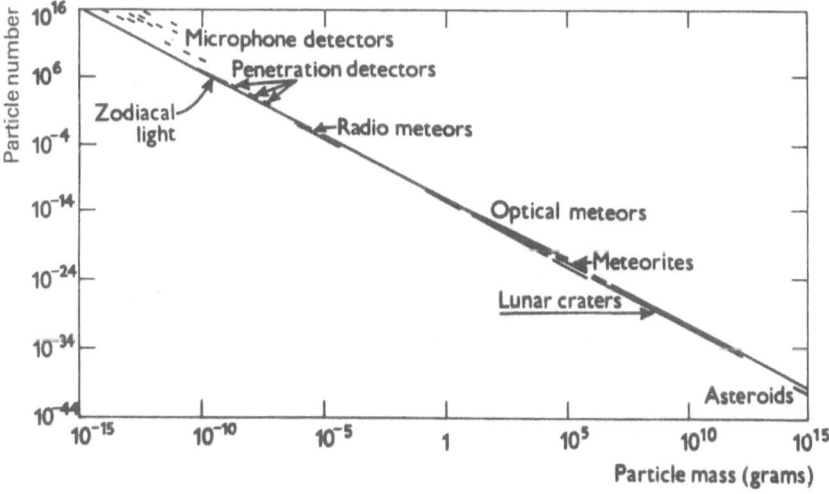

Fig. 4.10. Meteoritic complex. Mass of particles is on abscissae axis, particle number in a volume of $(100 \text{ m})^3$ is on the ordinate axis (after a model of J. K. Wall).

4.4.2. ASTEROIDS

In the mass spectrum of the planetary system asteroids represent a continuation of the meteoritic complex. Sometimes the asteroids are called minor planets or planetoids. In a telescope they appear as stars with apparent brightness between 7 and 21 magnitude, but they move in the sky – like planets. From comets they differ by absence of coma and tail.

It is generally accepted that the asteroids originated by accumulation of dust and gas in the primeval nebula from which the whole system was formed. Some asteroids crossing the Mars orbit are considered to be extinct nuclei of short-period comets (for example Icarus).

The brightness of some asteroids changes and their light curve points to their rotation and shape. Rotation periods lie between 3 and 19 h. As the light curve of the asteroids repeats over many periods, it follows that they rotate around one axis and do not wobble like spinning tops. Some asteroids are cigar-shaped like Hector which is about 100 km long and 40 km wide. The apparent diameter of Ceres (0.6″), Pallas (0.6″), Juno (0.2″) and Vesta (0.4″) was measured directly during their opposition. From known distances the size of the four asteroids was determined (radii of 380 km, 240 km, 100 km, and 240 km).

The color of asteroids is, in general, brownish-grey, similar to the color of the Moon and Mercury. It has been suggested also that the surface of asteroids (Figure 4.11) may be covered with dusty regolith. Probable density is about 3.5 g cm^{-3} (mean density of the Moon is 3.34 g cm^{-3}) so that the mass of Ceres would be about 10^{24} g;

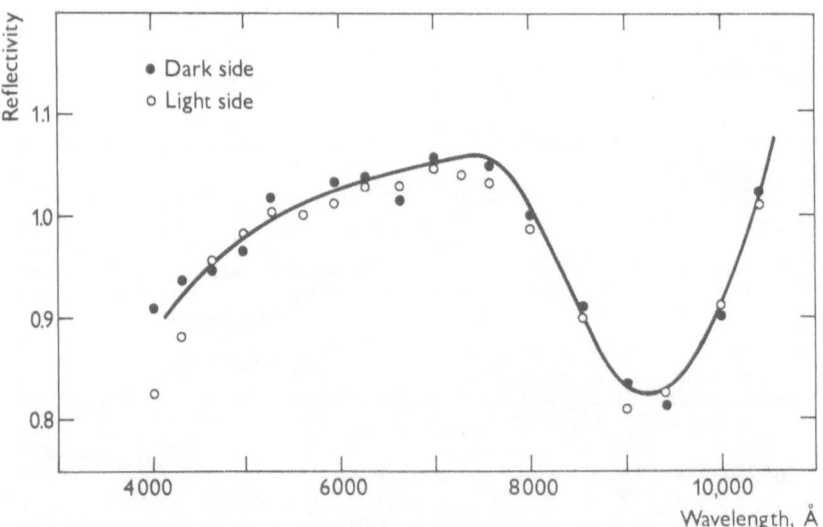

Fig. 4.11.   Relative reflectivity of the asteroid Vesta for different wavelengths. Deep minimum exists at about 9000 Å – indicating a high absorptivity of the asteroid surface at this wavelength (MIT and Mt Wilson Observatory).

it is the largest asteroid. On the other side of the mass range of asteroids is Adonis with a radius of 0.15 km and mass about 5×10^{13} g.

The number of asteroids registered in the Ephemeris of Minor Planets is over 1800, but the number of all observed asteroids is several thousands. The total number of asteroids brighter than 21 magnitude (absolute brightness is defined for one AU from the Sun and from the Earth and for full phase) or with a diameter $\geqslant 1.6$ km has been estimated to be 4.8×10^5. The number of asteroids increases as their mass decreases (compare Figure 4.12). The total mass of all the asteroids is about 3×10^{24} g, which is 0.4×10^{-3} of the Earth's mass and 10^5 times the mass of the meteoritic complex.

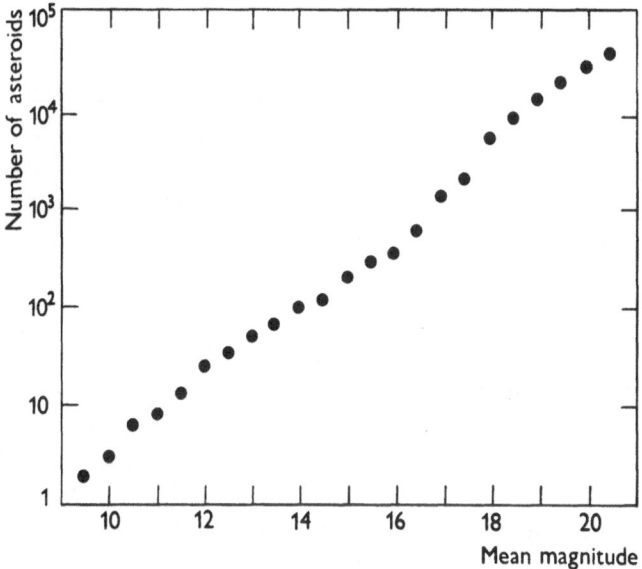

Fig. 4.12. Number of asteroids in the asteroid ring. Abscissa is mean magnitude for opposition. Each point shows the total number of asteroids of the corresponding magnitude and brighter (after van Houten).

The large majority of asteroids moves between the orbits of Mars and Jupiter (asteroidal belt). Some asteroids cross the Martian orbit (Mars crossers). More than 15 have been discovered which cross the Earth's orbit. Since none of these objects have been rediscovered, there must be many more of them. It is also possible that some asteroids cross the Jupiter orbit (e.g. Hidalgo) but they are difficult to discover due to their large distances from the Sun and the Earth.

The asteroids and their fragments – meteorites – are important for understanding the primitive stage of evolution of the planetary system. They are made of primeval material of the solar system and were less affected by later evolution than Moon and Earth. Some satellites of planets (e.g. Phobos and Deimos) are asteroid-like bodies and their genetic relation to asteroids remains a problem. The desire to learn the properties of the primeval material of the solar system may lead to asteroid missions

– it may be a flyby, an automated orbiter or a rendezvous for reconnaissance of the asteroidal surface. A sample-return mission would bring some material for detail laboratory investigations. There are no technical difficulties for such unmanned mission to asteroids – in particular to the near asteroids like Eros or Geographos.

4.4.3. STRUCTURE OF COMETS

None of the bodies in our planetary system undergoes such spectacular changes as comets. It is a consequence of their structure, composition and also of their elongated orbits.

(a) *Mass of Comets*

The masses of the brightest and heaviest comets are four to five orders of magnitude smaller than masses of the heaviest asteroids. The Halley comet has a mass about 2.5×10^{19} g while the mass of the asteroid Ceres is $\sim 10^{24}$ g. On the other side, the faint comet Wirtanen (1967.9) had a mass of $\sim 10^{15}$ g, so that it is about twenty times more massive than the asteroid Adonis (5×10^{13} g). The estimated number of all comets belonging to the solar system is 2.5×10^{6}.

(b) *Cometary Nucleus*

At large distances from the Sun a comet appears in a telescope as a starlike point. It is the cometary nucleus without any envelope. The nucleus probably represent a large dirty snowball. In its center is a solid core (~ 1 km) and the core is surrounded by icy or snowy volatile material, in which many dust particles are embedded (therefore a 'dirty snowball'). The radius of the nucleus is roughly 10 km.

The icy or snowy conglomerate (called nucleus) of volatile substances is the source of the transient phenomena like expanding coma, plasma tail and dust tail, bright outbursts or even splitting and other forms of cometary activity (Figure 4.13).

The size, composition, nature and shape of the compact solid core in cometary nuclei have not yet been determined. Even its existence itself is sometimes questioned.

(c) *Coma*

When the nucleus approaches the interior part of the solar system to distances $\leqslant 3$ AU from the Sun, the solar radiation heats the soft icy surfaces of the nucleus. Its volatile material sublimates and evaporates. The evaporated atoms, molecules and ions (such as O, Na, C_2, C_3, CN, CH, NH, NH_2, N_2^+, OH^+, CO^+, CO_2^+, CH^+) produce the spectacular coma for a transient time. The evaporating gases drag away icy grains, dust grains with an icy mantle or bare dust grains from the volatile surface of the nucleus.

Coma is a sphere of gas and dust around the nucleus, expanding in all directions. It is lost to interplanetary space with terminal velocities 0.6 to 0.8 km s^{-1}. Around the visible coma there is a huge *hydrogen halo*. It expands considerably faster, about

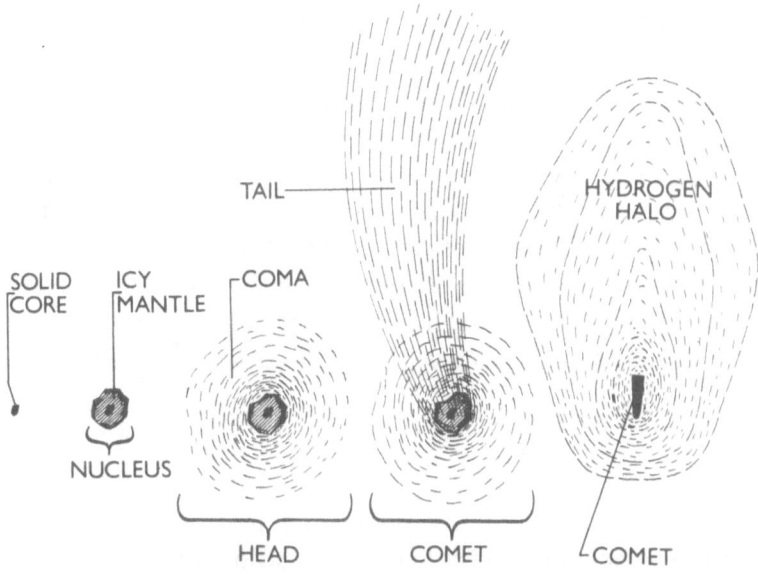

Fig. 4.13. Structure of comets. The hydrogen halo is much larger than the visible comet (dark feature in the last figure).

$8 \, km \, s^{-1}$, as has been deduced from $L\alpha$ isophotes and the $L\alpha$ profiles. This and other spectroscopic observations indicate that the production rate of OH, O and H from the whole nucleus is about 10^{30} atoms (i.e. 1 t) per second, and depends on the distance of the comet from the Sun. Since the production of gases lasts for about 10^6 s, the nucleus loses $\sim 10^6$ t during one appearance of the comet.

The abundant molecules observed in cometary spectra, i.e. C_2 and CN, are produced at a rate of 10^{28} molecules per second. This implies that the common parent molecule is water and that the water-ice or water-snow is the principal constituent of the volatile fraction of the cometary nucleus. The other chemical constituents cannot represent more than a few percent, because otherwise the coma could not develop at a distance of 3 AU. The size of the coma changes with distance from the Sun and it may be from 10^4 to 10^6 km in diameter.

The expansion of the solid particles in the coma is slightly slower (velocities 0.3 to 0.6 km s^{-1}) than the expansion of molecules and ions. The **dust production rate** from the nucleus is one or two orders of magnitude lower than that of the gas and plasma. But in some very dusty comets both production rates are comparable.

(d) *Tail of Comets*

At still smaller distances from the Sun ($\leqslant 1.5 \, AU$) a tail is formed. Molecules and ions are expelled from the coma by pressure of electromagnetic and corpuscular radiation of the Sun. Acceleration of the material in cometary tails in terms of the solar gravity is about 50 to 150 times higher but may be much greater. Fine streamers observed in tails are often twisted: the motion of the tail plasma (OH^+, N_2^+, CH^+, etc) is guided by magnetic lines of force slightly twisted and stretched behind the coma over distances of 10^7 to 10^8 km. The tail material is lost from the nucleus forever. Old short-period comets (e.g. Encke comet) have short inconspicuous tails.

The electromagnetic interactions are of main importance for the structure and evolution of comets: the nucleus is held together more by intermolecular elec-tromagnetic forces than by self-gravitation, interaction of the solar radiation with the volatile material of the nucleus, acceleration of the tail material and its motion along the magnetic lines of force – these are some examples. Gravitation determines the orbit of the comet around the Sun, disturbances of its motion by planets and formation of an anomalous tail consisting of larger particles and directed towards the Sun.

(e) Mass Losses

The gas production from the cometary nucleus may be ~ 10 t s^{-1}. The dust emission rate may be comparable, being a few tons per second in very dusty comets. However in some comets it is one or two orders of magnitude lower.

Comets are associated with meteor showers. The shower meteoroids contain more volatile elements than the non-shower meteorites. The dust lost by comets is of the right order of magnitude to replenish the meteoritic complex. The comets are supposed to arrive from an extended Oort's cometary cloud in instellar space behind the outskirts of the planetary system. They transport material from the nearby interstellar space into the interior of the planetary system.

4.4.4. SATELLITES OF PLANETS

There are 32 known satellites accompanying six planets in our solar system. The Moon has a mass 81 times smaller than the Earth's mass and 73 times larger than that of the largest asteroid Ceres. There are three satellites in our planetary system with mass about twice as large as the Moon (Ganymede of Jupiter, Titan of Saturn and Triton of Neptune); three satellites (Io, Europa and Callisto of Jupiter) have comparable mass with our Moon. Eight satellites have a mass similar to that of the largest asteroids and the remaining 17 satellites are small, irregular bodies, similar to smaller asteroids. The largest satellites are more similar to terrestrial planets (e.g., Mercury) than to the smallest satellites. The recent flyby mission to Mercury has shown a striking resemblance of the innermost planet to our Moon.

4.4.4.1. Moon

Our knowledge of structure of the Moon has considerably increased recently, thanks to automated stations and measuring instruments on its surface but especially thanks to direct exploration *in situ* by men.

(a) *Lunar Globe*

The mass of the Moon is 73×10^{24} g. Its mean radius is 1738 km, far above the critical size (about 400 km) at which a solid body is crushed by self-gravitation to spherical shape with minimum potential energy. The Moon is very approximately spherical. Its mean density 3.34 g cm^{-3} is close to the density of rocks collected on the lunar surface by astronauts. This implies that there is no significant increase of density towards the Moon center. It also means that it consists mainly of silicate rocks.

(b) *Internal Structure*

Information on internal structure of our satellite is being received from moonquakes both natural and man-made. Since 1969 the seismometers installed on the lunar surface have recorded more than one thousand individual moonquakes. The path and velocity of seismic waves propagating in the lunar globe bear important information on its structure, though the seismic energy of the Moon is 9 to 10 orders of magnitude less than the seismic energy of the Earth. The largest moonquakes are of the magnitude 2 to 3 (on the Richter scale), which is just a sensitivity for man. On the whole, the Moon is tectonically a very quiet body.

The propagation of waves in the lunar globe indicates that to a depth of about 900 km (lunar mantle) it behaves like an elastic solid rock in which both compressional and shear waves propagate. Lunar seismographs have recorded no shear waves coming from depths below 900 km which indicates that the central part of the Moon behaves like a plastic substance.

(c) *Lunar Soil (Regolith)*

The solid lunar globe is covered by a layer of loose broken rock material called regolith. Its depth, as measured by the reflection of decametric waves from terrestrial radars, is 50 to 100 m. It consists of loosely packed dust, gravel and boulders – as may be seen in photographs of abrupt slopes, e.g., of the Hadley rille. Study of drill cores indicates that the size of particles increases with depth: mean diameter at surface is 60 μm and at a depth of 240 cm it is 110 μm. Also their density increases from 1.3 to 2.2 g cm^{-3} at the 240 cm depth. Such heterogeneous layer scatters intensely seismic waves nearly without losses. This is a reason why lunar seismograms are markedly different from recordings of terrestrial earthquakes.

The lunar soil (regolith) is covered with a thin dust layer. The footprints of astronauts have a depth of 3 mm to 5 cm and the tracks of the rover wheels are 8 to 15 cm deep. Particles of all sizes from ~1 μm up to ~10 m boulders are found on the surface. The regolith layer with its dust cover is a transition layer between the lunar globe and cosmic space.

(d) *Age and Changes*

The regolith has been shaped and turned over from both sides: by internal activity of the globe and by fluxes of photons, solar wind, cosmic radiation, by impacts of meteoroids, asteroids and comets from outer space. The physical and chemical properties of the lunar soil and its present forms (craters, rilles, rays, . . .) are results of activity of external and internal factors, an activity which began about 4.6×10^9 yr ago. The youngest crystalline rocks are 3.2×10^9 yr old which is not much less than the age of the oldest terrestrial rocks ($\sim 3.5 \times 10^9$ yr) found in South Africa and in Canada. Lunar breccias and dust are up to 4.6×10^9 yr old. Mean age of meteorites (i.e., solidification age) is 4.5×10^9 yr. All these numbers indicate the time of the formation of our planetary system. The lunar surface reflects the oldest history of the system.

4.4.4.2. *Natural Satellites*

Next to the Moon, Phobos and Deimos of Mars, the four Galilean satellites of Jupiter and Titan of Saturn are the most studied natural satellites.

(a) *Size*

Only the Galilean satellites and Titan are near and large enough to enable us to measure directly their radii. The size of Phobos and Deimos have been determined precisely from close-range photographs taken by the Mariner spacecraft. Both are irregular and their mean radii are 10.9 km and 5.7 km. Mean radius is the radius of a sphere whose volume equals that of the satellite. Sizes of all the other satellites are determined from their photometric magnitude and albedo.

(b) *Mass*

Mass could be determined only for the large satellites (with mass $> 10^{22}$ g) whose gravitational perturbations upon each other can be observed from the Earth. The mass of Titan is, for example, known (1.9 times the lunar mass) with a precision of 1% from its influence on the motion of Hyperion, its nearest neighbor in the Saturn family. The Doppler tracking of a spacecraft passing through a satellite system is a modern possibility for mass determination of satellites. The mass of Io (9×10^{25} g) was determined by Doppler tracking of Pioneer 10.

There are together ten satellites (excluding the Moon) with known masses and ten with known radii – but only six with both quantities determined (four Galilean satellites and two satellites of Saturn). In all the six cases the mean densities are smaller than for the terrestrial planets. The density of Galilean satellites decreases with their distance from the planet (i.e., $3.2 \, \mathrm{g \, cm^{-3}}$, $3.0 \, \mathrm{g \, cm^{-3}}$, $2.0 \, \mathrm{g \, cm^{-3}}$ and $1.4 \, \mathrm{g \, cm^{-3}}$). The mean densities of satellites show a tendency to decreases with distance from the Sun: Moon $3.34 \, \mathrm{g \, cm^{-3}}$, Galilean satellites $1.4 \, \mathrm{g \, cm^{-3}}$ to $3.2 \, \mathrm{g \, cm^{-3}}$, Saturn satellites $1.1 \, \mathrm{g \, cm^{-3}}$ to $1.4 \, \mathrm{g \, cm^{-3}}$.

(c) *Rocky Satellites*

The satellites of the inner planets (Earth and Mars) are rock-like. They are formed from non-volatile substances, because unprotected frozen volatiles on the surfaces of bodies inside the orbit of Jupiter would vaporize.

Phobos and Deimos are well consolidated solid rocks. Aggregates of small grains, held together by self-gravitation would not survive large impacts suffered by the satellites – impacts that excavated craters up to 5 km in diameter. The surface of Phobos and Deimos is covered with porous dust layer as follows from very fast change of surface temperature during an eclipse of Phobos, measured by IR radiometer onboard Mariner 9. The low thermal conductivity excludes a solid rocky surface.

(d) *Icy Satellites*

The unprotected frozen volatiles can persist on the satellites of Jupiter and those of the more distant planets. To the non-volatile substances ice-forming compounds are added which decrease the mean density and increase the albedo of the satellites.

The computed models of satellites show that for average densities $>3.0 \, g \, cm^{-3}$ satellites are composed of rocks, those with densities 1.5 to $3.0 \, g \, cm^{-3}$ from rock, water ice and possibly NH_3 and CH_4 hydrates; for much smaller densities than $1.5 \, g \, cm^{-3}$ there should be substantial quantities of pure CH_4. Thus Jupiter satellites consist of rock and ice only, while some Saturn satellites contain also CH_4 ice in addition to the rocky component and water ice.

All satellites contain rocky materials presumably with some radioactive sources of heat, like K, U and Th in chondrites. The heat may have serious consequences for the internal structure of the icy satellites. Their heated interior probably melts and according to the computed models it should be mostly liquid with a rocky nucleus and icy crust. The comparison of IR spectra of some satellites (e.g. Ganymede, Europa) with laboratory spectra indicates that H_2O ice may be the main constituent of their surfaces.

(e) *Atmospheres and Ionospheres of Satellites*

Atmospheres have been suspected on some satellites of Jupiter and of the more distant planets. An ionosphere has been measured on Jupiter satellite Io by occultation of the spacecraft Pioneer 10. The occultation was measured on 2200 MHz and gave electron-density profiles of the Io ionosphere. Its maximum electron density (6×10^4 electrons cm^{-3}) at an altitude of about 100 km implies a density 10^{10} to 10^{12} molecules per cm^3 at the satellite's surface. The UV photometer onboard the spacecraft recorded in $L\alpha$ a bright halo of H around the satellite. The bulk of atmosphere consists of NH_3 and its dissociation products N_2 and H_2.

The largest satellite of Saturn and probably also the largest satellite in the whole solar system, viz. Titan, has a sufficient gravity ($\approx 140 \, cm \, s^{-2}$) and a great distance from the Sun to possess an atmosphere. Molecules CH_4 and H_2 are detected in its spectrum. The CH_4 is expected to condense into thick clouds in the Titan's atmosphere for temperatures below 80 K. A strong evidence for the clouds comes also from anomalously low albedo in the UV radiation (i.e., absence of Rayleigh scattering).

4.4.5. PLANETS

The largest solid bodies are planets. It is generally accepted that Jupiter's mass is only slightly less than the mass of the lightest stars. A body with a larger mass (e.g. 0.06 $M_\odot$) would be self-luminous – it would be a star (Section 5.4.2); its self-gravitation is namely a sufficient source for the body to radiate – even if dimly.

(a) *Mass and Density of Planets*

According to their mass and density the nine planets of our planetary system may be subdivided into four groups: (1) lesser planets with mass one order smaller than the Earth mass (Mercury 0.06 M_E, Mars 0.11 M_E); (2) Earth and Venus with masses nearly equal (1.0 and 0.82 M_E) and mean densities 5.5 g cm^{-3} and 5.2 g cm^{-3}; (3) Neptune and Uranus with masses one order larger (17 M_E and 15 M_E); (4) the largest planets with mass two order of magnitude larger (Jupiter 318 M_E and Saturn 95 M_E) and with the smallest densities (1.4 g cm^{-3}, respectively, 0.7 g cm^{-3}). Pluto, considered to be a lesser planet (with mass 0.2 to 0.8 M_E and radius 0.5 R_E to 0.6 R_E) is poorly known and practically nothing can be said about the most distant of the nine planets. Due to the great lack of information it is usually left aside in planetary astronomy.

Earth, Venus, Mars and Mercury are often called terrestrial planets or inner planets, while Jupiter, Saturn, Uranus and Neptune are called giant planets, Jovian planets or outer planets.

(b) *Structure of Terrestrial Planets*

The structure and evolution of the terrestrial planets is governed by their small mass, large densities and small distances from the Sun (Section 5.5). They differ from the giant planets with large masses, small densities and large distances from the Sun both by structure and evolution. Many conclusions about the structure of terrestrial planets are based on the knowledge of the Earth's interior, deduced primarily from the propagation of seismic waves.

The Earth consists of two main parts: a core with a radius 3500 km and a mantle which surrounds the core. The S-waves do not propagate through the core. It is therefore supposed to be liquid (plastic), though the properties of material at pressures $\sim 10^6$ atm. and temperatures ~ 4000 K are not completely understood.

The interface between the core and the mantle is 2900 km deep and there is a sudden density change from 5.7 g cm^{-3} to 9.4 g cm^{-3} at a pressure of 1.4×10^6 atm. The high density of the core is due to its chemical composition: it is mainly metallic (with Fe and Ni as the principal constituents) with admixtures of molten silicates.

The mantle is solid and transmits both types of seismic waves – compressional P-waves and shear S-waves. The density decreases outwards from 5.7 g cm^{-3} at the depth of 2900 km to 2.6 g cm^{-3} in the crust (lithosphere), i.e. the uppermost layer about 30 km thick. In the crust and in the upper part of the mantle the pressures are relatively low and the properties of the material there are determined by a great variety of minerals and rocks. The electromagnetic forces between ions of crystals (lattice forces) are considerably greater than the external hydrostatic pressure caused by gravity. The forces determine the geometry of the lattice, the density, elasticity, thermal and electrical properties of the crystals. At the lower parts of the mantle the lattice forces yield to the hydrostatic pressure. Gravitation prevails over electromagnetic interactions.

No seismographs have so far been placed at the surface of other terrestrial planets to provide direct information about their interior. To compute a model of a planet its mass, radius, chemical composition, rotational velocity, and oblateness are to be known. The theory of internal constitution of planets is based on differential equations which express hydrostatic equilibrium (i.e., change of pressure dP/dr) and temperature gradient (change of the temperature dT/dr with distance from the center r). A few supplementary relations between the quantities involved (such as equation of state, energy production by radioactive elements and gravitational contraction, heat conductivity which contrary to stellar interior is mainly due to conduction, phase transitions with a discontinuity in density, variation of melting temperature with pressure) are needed for numerical integration of the differential equations. The integration proceeds from the surface of the planet radially inward.

The densities and temperatures in the interior of the planets are relatively low if compared with the stellar interior, and the behavior of matter is therefore complicated. The equation of state and other relations are not precise and the chemical composition at different depths moreover is uncertain. The resulting models giving distribution of density, temperature and pressure in the interior of the planet may differ from one author to another. This explains the surprising fact that the internal constitution of a very distant star may be better known than the interior of a nearby planet in our system. The properties of matter at high temperatures and densities are not so complicated (Chapter 3), and are sufficiently well understood so that the interior of most stars is satisfactorily explained.

(c) *Structure of Giant Planets*

The giant planets have low average densities and they consist primarily of light elements, with H and He as major constituents. Saturn with its density $0.7 \, \text{g cm}^{-3}$ cannot consist of much rocky materials (except the rocky core) and its internal structure will be substantially different from the rocky terrestrial planets. Special attention among the giant planets has been paid to the structure of Jupiter and Figure 4.14 represents a typical model.

The interior of Jupiter must produce heat, because the planet emits 2.7 times more energy (as IR radiation) than it receives from the Sun. The source of the surplus energy is probably self-gravitation. If the chemical composition of Jupiter is similar to the solar composition (Figure 4.2) then the abundance of radioactive elements is insufficient to explain the extra heat. For thermonuclear reactions the temperatures are too low. A probable source of the Jupiter IR excess which represents 10^{33} erg per year, is a gradual gravitational shrinkage of the planet. The same source could also

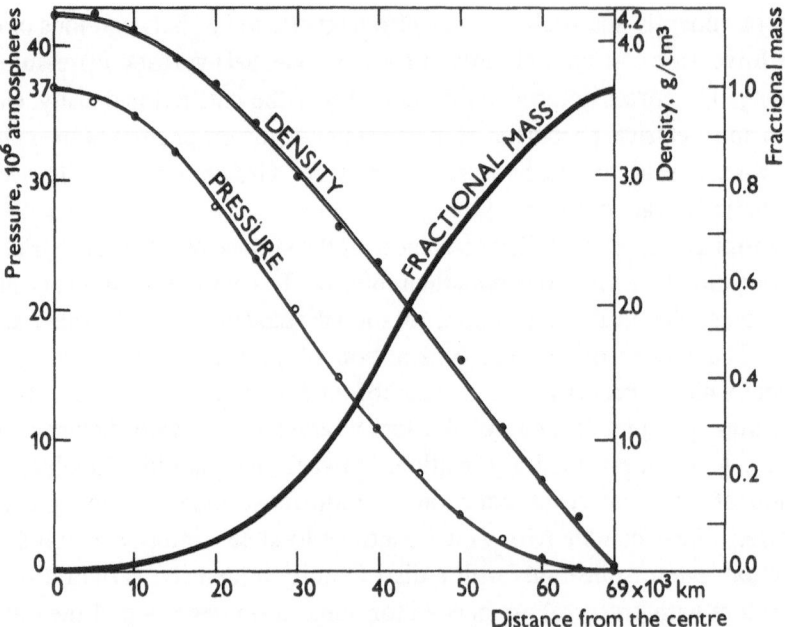

Fig. 4.14. Structure of Jupiter (according to Hubbard, De Marcus, Peebles). Pressure, density and fractional mass are given for each distance from the planet's center.

explain Saturn's radiation which is 2.4 times the solar energy received. The excess of Uranus radiation is also explained by gradual gravitational shrinkage. The mechanism of energy liberation consists in a radial growth of the dense metallic hydrogen core at the expense of the overlying less dense H_2. A growth of 1 mm yr^{-1} would release the correct amount of energy.

Another observable property of the Jovian interior is the planet's magnetosphere, with its north magnetic pole near the north pole of rotation (which is opposite to the situation on Earth). The magnetosphere is responsible for the decametric and decimeter radiation of the planet. *In situ* measurements by Pioneer 10 confirmed that existence of the magnetosphere and trapped particles which emit the radiation.

In general, the magnetic field of planets is one of the few properties which is directly measurable and gives information on the planetary interior. At least a dozen theories have already been proposed for the origin of planetary magnetism, but the theory of hydromagnetic dynamo seems now generally accepted. The magnetic field in the planet is enhanced and maintained by energy of convection in liquid conducting layers. Jupiter (and Saturn) have internal heat sources which may drive an intense convection (convectionally driven dynamo). The sources of the Earth are insufficient to account for the observed geomagnetic field and the main driving force for the terrestrial dynamo is the Earth's precession (precessionally driven dynamo).

(d) *Planetary Atmospheres*

The gaseous envelope of planets – atmospheres – form a transition between their solid or liquid surface and the surrounding interplanetary space. They belong naturally to the structure of planets and play an important role in their energy balance. The total mass of an atmosphere, its chemical composition, distribution of densities and temperatures with altitude – all depend on the mass of the planet, its distance from the Sun and its evolution.

It is now generally accepted that the Jovian atmosphere, like the Sun, preserves the original chemical composition of the primitive nebula (Section 5.5). If one calculates the radius of a cold sphere with the mass of Jupiter or Saturn composed of H and He (with relative solar abundances), one obtains the actual radius of the planets. This is a strong argument in favor of the assumption that their overall composition is similar to that of the Sun.

Photometric observations of stellar occultations by Jupiter and analysis of spectra taken across the planet's disk give information on the structure of the Jovian atmosphere. Temperature and pressure are increasing with depth. There are at least two cloud layers with a layer of clear atmosphere in between. The upper thin semi-transparent layer consists probably of solid NH_3 whereas the lower thick cloud deck may consist of water ice and NH_4SH. The structure of the Jovian atmosphere is inhomogeneous (Figure 4.15) and variable, as may be seen from detailed photographs of the planet. Besides the hydrodynamical activity of clouds and the Red Spot,

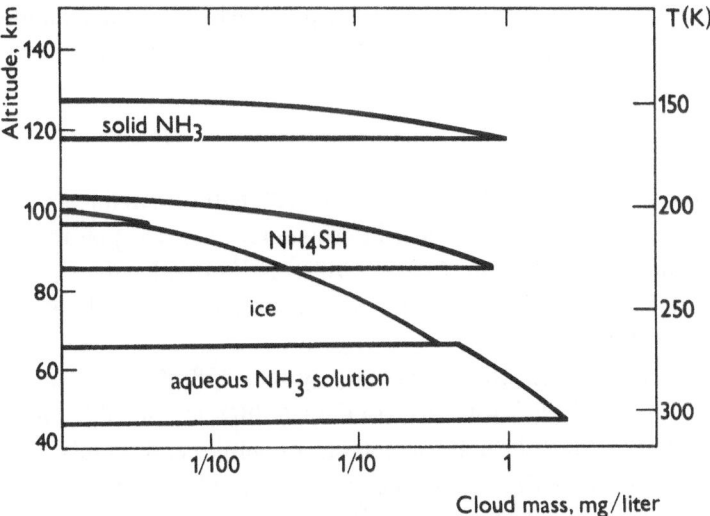

Fig. 4.15.   Clouds in the Jupiter atmosphere (Lewis).

there is also an intense chemical activity. The changes in coloration have been interpreted by synthesis of complex organic molecules like aminoacids. The absorption features in the Jupiter spectrum at 2600 Å seem to be due to adenine, which is a basic constituent of DNA and RNA. If this idea corresponds to reality, then the Jupiter atmosphere would be the place of intense prebiological activity in which molecules important for evolution of life are synthesized. When in laboratory experiments CH_4, NH_3, and H_2O (which after H and He are the major constituents of Jovian atmosphere) were irradiated by electrons, adenine among others has been formed. There are fluxes of high energy electrons in the Jovian magnetosphere necessary for the synthesis.

The atmospheres of terrestrial planets are secondary formations. Their chemical composition is completely different from the atmosphere of Jupiter and from the 'universal composition' of the planetary system (Figure 4.2). A very thin atmosphere has been recently detected on Mercury with He being its main constituent. It may either be captured together with Ne from the solar wind or partly released with Ar by decay of radioactive elements in the crust of the planet.

The tenuous atmosphere of Mars and the hot heavy atmosphere of Venus are composed mainly of CO_2. The CO_2 in the Venus atmosphere is about 10^5 times more abundant than in the Earth's atmosphere, though the total amount of CO_2 on both planets need not be different. On our Earth it is mostly bound in its crust in form of carbonates, so that its abundance in the atmosphere is low. Nevertheless it plays an important role in thermal equilibrium of our planet, together with H_2O and O_3. These gases absorb very efficiently the IR radiation emitted from the Earth's surface. The absorbed energy is re-radiated, partly backwards towards the surface, and increases the ground temperature by about 30 K. This phenomenon – transparency of atmosphere for light and opacity for IR radiation – is called the greenhouse effect. It is especially important in the Venus atmosphere which contains mainly CO_2. It explains the high temperatures measured on the planet's surface.

4.4.6. SOLAR SYSTEM

The system is composed of the Sun, nine planets, 32 natural satellites, a ring system, asteroids, comets, meteoritic complex, interplanetary gas and plasma. Most of its mass is concentrated in the Sun ($M_\odot = 1.989 \times 10^{33}$ g) and only a small fraction $M_\odot/743 = 448\ M_E$ is distributed among other members of the solar system (M_E is the Earth mass 5.976×10^{27} g).

From the mass of the whole planetary system $448\ M_E$ most, i.e. $447.8\ M_E$, belongs to planets; $0.12\ M_E$ is total mass of the satellites, about $0.0003\ M_E$ belong to asteroids, while $10^{-9}\ M_E$ is the mass of meteoritic complex and comets. The numbers imply that most of the matter in the solar system is in plasma state and only about 0.1% is solid, gas or liquid. There is no degenerate of neutron matter within the confines of the system. Relativistic nuclei and electrons of galactic cosmic rays, solar cosmic rays, sub-cosmic radiation and solar wind are more important by their energetic interactions with members of the system than by their quite negligible mass.

The principal force binding the solar system together is gravitation. Satellites and the ring-system are bound to their planets, planets in turn and all the other members of the system move in the gravitational field of the Sun. Their motions and distances exhibit some regularities. The orbits of planets are quasi-circular – i.e. ellipses with small eccentricities. They lie almost in the same plane – called the ecliptic. The planets move in their orbits in the direction of the solar rotation. The angular momentum of the solar system is nearly parallel with the rotation axis of the Sun. Most of the angular momentum of the solar system is concentrated in planets $(3 \times 10^{50} \, \text{g cm}^2 \, \text{s}^{-1}$ compared with the $1.6 \times 10^{48} \, \text{g cm}^2 \, \text{s}^{-1}$ of the Sun), though the mass of planets is much smaller than the mass of the Sun. The distances of planets from the Sun approximately fit an empirical rule known as the Titius–Bode law

$$0.4 + 0.3 \times 2^n \tag{4.6}$$

where n is $-\infty$ for Mercury, 0 for Venus, 1 for Earth, 2 for Mars, 3 for asteroids, etc. Other forms are sometimes used. The law (or better to say, the rule) led to discovery of asteroidal belt in the place where, according to expression (4.6), a planet should appear.

An important feature of the solar system is its chemical composition: the inner, terrestrial planets and their satellites consist of rocky silicate material, while the outer members – i.e., giant planets, their satellites and comets are made of light icy materials. It is mainly a question of absence or presence of light elements (H, He) and a result of evolution. On the other hand the heavier elements seem to have the same 'universal' abundances in all members of the system, the same as in the solar atmosphere and in the primitive nebula (Figure 4.2). All the present properties, not only the element abundances, of the present solar system have been inherited from the primitive nebula. It is the aim of planetary cosmogony to explain how it happened (Section 5.5).

4.5. Structure of the Stars

Stars represent fundamental structural units in the hierarchy of the universe, in the sense that most of the observable matter in the universe is concentrated in them and that the higher structural levels (clusters, galaxies) are systems of stars. They are the most massive bodies in the universe.

(a) Interactions in Stars

The principal force binding all particles of a star together is their mutual gravitational attraction (self-gravitation). It determines the structure of a star and, according to the virial theorem, gravitational energy is converted into internal energy of the stellar interior. As such, the weakest interaction of elementary particles is the decisive force for the whole existence of a star, from its birth by gravitational contraction to its end by gravitational collapse.

The other three interactions participate in the structure and evolution of stars. Thermonuclear transformation of protons into He nuclei and heavier nuclides is a consequence of strong interactions between nucleons at distances $\approx 10^{-13}$ cm. The strong interactions are the main energy source of a star during most of its life. They produce practically all nuclei for the elements of the Mendeleev periodic table (nucleosynthesis). On a much smaller scale, strong interactions temporarily act also in stellar atmospheres, where fluxes of high energy protons are accelerated to relativistic energies by magnetic fields.

The weak interactions have a still much shorter range than the strong interactions ($<10^{-15}$). The decay of nucleons in Equations (2.19) and (2.20) in nuclei constitutes a weak process very important for the life of stars. Without weak interactions there would be no decay of protons into neutrons and no element could be synthesized from protons. Nuclei composed only from protons do not and cannot exist. The nucleosynthesis is produced by strong interactions, but it would not proceed without intervening weak decays, when a synthesized nucleus emits excess energy in the form of positron (e^{+}) and neutrino (ν_e) and thus becomes more stable.

The electrostatic repulsion between colliding nuclei acts on long distances. It regulates the rate of thermonuclear reactions (by the Coulomb barrier of repulsive forces). If the electric repulsion would stop, the nuclei of the whole star would immediately strongly interact with liberation of a huge amount of energy. A large majority of observed stars is plasma bodies which implies that also other elementary processes (collisions of ions and electrons, their Larmor gyration in magnetic fields, emission and absorption of photons) are governed by electromagnetic interactions.

One may conclude this item by saying that all four interactions between elementary particles are involved in the structure and evolution of normal (i.e. plasmatic) stars.

(b) *Stellar Interior and Stellar Atmosphere*

From the observational point of view, a star consists of two main parts: the interior and the atmosphere. All radiation of a star is emitted from its atmosphere and no photon leaves stellar interior directly for interstellar space. In other words: atmosphere is the visible part of a star, while its interior is not observable (except, possibly, through the flux of neutrinos – see Section 2.3.8).

Stellar atmosphere represents a very small fraction of the total stellar mass (i.e. roughly about 10^{-10}). Practically all matter of a star is hidden in its interior. The interior thus represents a storage of mass and energy under different forms such as photons, thermal, potential, rotational, etc. The stellar interior is a quasi-equilibrium concentration of mass and energy, surrounded by an inter-stellar medium in complete non-equilibrium.

A stellar atmosphere represents a transition zone between the stellar interior and the interstellar medium characterized by propagation of photons and particles. The transient regime from storage to propagation appears as radiation fluxes and particle streams in the stellar atmosphere.

(c) *Observed Parameters of Stars*

The main parameters of a star are: mass M (or number of baryons), luminosity L (i.e., energy radiated per second), surface temperature T (characterizing spectral distribution or energy output from 1 cm^2 of the stellar surface), radius R and chemical composition. Each star may be represented by a point in the three dimensional space (M, L, T). (The fourth parameter – chemical composition – is less important). The points representing actual stars occupy only certain regions of this space. An arbitrarily chosen point (i.e. set of M, L, T values) may not correspond to an actual star. Usually two-dimensional projections on the (T, L)-plane and the (M, L)-plane are used, i.e. the Hertzsprung-Russell diagram $L(T)$ and the mass-luminosity relation $L(M)$. Both relations between the observed stellar parameters represent a corner-stone for the theory of stellar structure.

(d) Computed Models of Stellar Interior

To compute a model of stellar interior means to determine for each distance r from the star center the physical parameters: pressure $P(r)$, density $\varrho(r)$, temperature $T(r)$, mass inside the sphere with radius r i.e. M_r, luminosity of the sphere L_r, energy production per gramm $\varepsilon(r)$ and opacity of the material $\varkappa(r)$. (For rotating stars the parameters depend also on distance from this equator.)

Parameters ϱ, T, M_r and L_r are found by solving a system of differential equations which express:

(a) the hydrostatic equilibrium between the inward directed gravitation and outward directed pressure of plasma

$$\frac{dP_r}{dr} = -\frac{GM_r \varrho_r}{r^2};$$
(4.7)

(b) the energy generation – relating the change of luminosity at distance r with the energy generation rate at that distance

$$\frac{dL_r}{dr} = 4\pi r^2 \varrho \varepsilon;$$
(4.8)

(c) the mass in a spherical shell of thickness dr

$$dM_r = 4\pi r^2 \varrho\, dr;$$
(4.9)

(d) the energy transfer by radiation

$$\frac{dT}{dr} = -\frac{3}{4ac}\frac{\varkappa\varrho}{T^3}\frac{L_r}{4\pi r^2} \quad \text{and}$$
(4.10)

(e) the energy transfer by convection

$$\frac{dT}{dr} = \left(1 - \frac{1}{\gamma}\right)\frac{T}{P}\frac{dP}{dr}.$$
(4.11)

Pressure P, opacity $\varkappa$ and energy generation rate ε are functions of density ϱ temperature T and chemical composition of the stellar plasma. Their determination is a purely physical problem.

Observed stellar parameters at the surface $(r = R)$, viz. $L_r = L$, $M_r = M$, $T_r = T_e$, chemical composition and some self-evident relations for the stellar centre (such as $M(0) = 0$, $L(0) = 0$) are boundary conditions for the solution of the equations.

Analytical solutions of the equations exist for simplified cases but are of little value. High-speed computers have to be used for calculation of realistic stellar models.

The structure of a star depends on its mass and chemical composition. If one constructs a series of models having different masses but the same chemical composition (uniform throughout the stars), the models lie on the main sequence in the HR diagram: massive models are on the upper left, light-weight models on the lower right. In the $L(M)$ graph the models fit the observed mass-luminosity relation.

The $L(T_e)$ and $L(M)$ functions for the main sequence may be deduced directly from the basic equations. For the known parameters of a star M, L, T_e and its chemical composition, there is a unique solution of the system of equations. Hence we can write

$$L_r = L_r(r, M, L, T_e, \text{chem. composition})$$
$$M_r = M_r(r, M, L, T_e, \text{chem. composition}) \tag{4.12}$$

and similar expressions for $T(r)$ and $\varrho(r)$. For the center of the star the relations (4.12) are

$$0 = L_r(0, M, L, T_e, \text{chem. composition})$$
$$0 = M_r(0, M, L, T_e, \text{chem. composition}). \tag{4.13}$$

These are necessary conditions for a star to exist. Not any arbitrarily chosen triad (M, L, T_e) will satisfy the relations (4.13). The main sequence of the Hertzsprung-Russell diagram $L = L(T_e, \text{chem. composition})$ and the mass-luminosity relation for the main sequence stars $L = L (M, \text{chem. composition})$ are equivalent to Equation (4.13).

The agreement of theory with observations implies that the physics behind the equations and in parameters $P, \varkappa$ and ε is in principle correct; the models will represent the structure of actual stars with a good approximation.

4.5.1. The Sun

Due to its nearness, the structure of the solar atmosphere is known in great detail – much better than that of any other star in the universe. The invisible interior (Figure 4.16) is studied by methods just described, viz. by numerical solution of Equations (4.7) to (4.11). There is some hope to directly probe the solar central core with the aid of neutrinos which should result from He synthesis; but the observations indicate considerably lower neutrino fluxes than those predicted by theory. This discrepancy remains open.

(a) *Energy Generation*

The matter in the solar interior is in a state of highly ionized plasma. The hottest and densest plasma is in the central core with temperatures about 13×10^6 K and

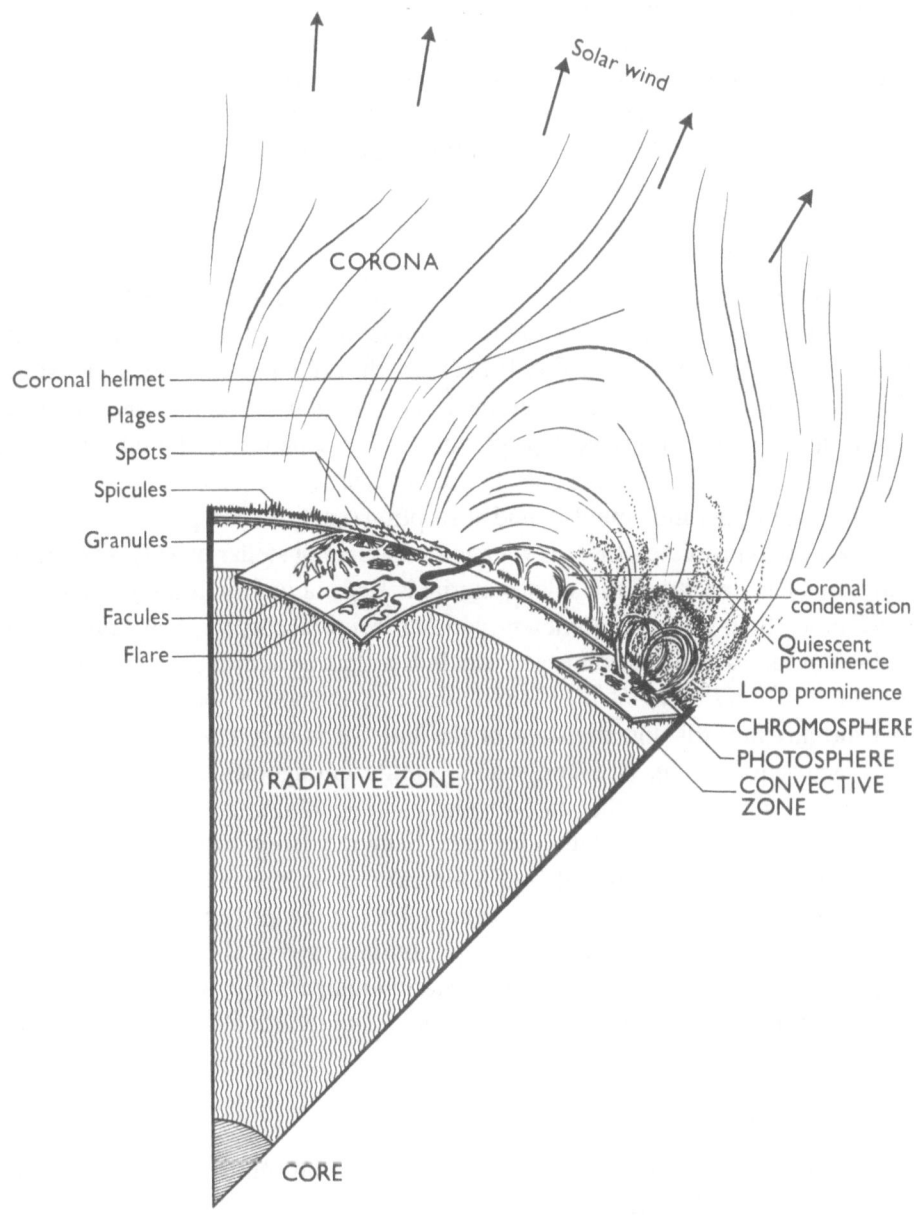

Fig. 4.16. Structure of the Sun is representative for many stars. Core, radiative zone and convective zone together are solar interior. Photosphere, chromosphere and corona are directly observable and are layers of the solar atmosphere. Regions with a strong magnetic field (i.e. active regions) are shown with different phenomena.

densities of 100 g cm^{-3}. At such conditions the H is transformed gradually into He – a process which is the energy source (ε in 4.8) of all the main sequence stars. In the interior of the Sun, four protons are fused into one α-particle by a sequence of reactions called the proton–proton chain (Table 2.3). The charge and mass differences (mass defect) are emitted in form of two positrons, two neutrinos and two gamma photons. The net energy generation is $26.2 \text{ MeV} = 4.2 \times 10^{-5}$ erg per one chain.

(b) *Energy Storage*

The generated energy is stored for a long time in the solar interior, with the exception of neutrinos. In the central core the radiation is in form of hard X-protons with Planckian distribution of 13×10^6 K. From there the radiation slowly leaks outwards, in the direction of the temperature gradient. The relation between the radiation flux L_r and the temperature gradient (Equation (4.10)) depends on the opacity of the stellar material in the radiative zone. The photons are absorbed by photoionization, free-free transitions and bound-bound absorption (i.e. line absorption). Besides the true absorption processes, there is yet another source of stellar opacity, namely the scattering. The opacity of the solar material is large, a photon has little chance to travel more than a few centimeters without being absorbed. The temperature gradient in such a small region is very small, less than a thousandth of a degree per centimetre ($10^7 \text{ K}/7 \times 10^{10}$ cm). The radiation flux outwards is only negligibly larger than the flux inwards and the state of the interior is very close to that of thermodynamic equilibrium. Thermodynamic equilibrium signifies that all the distribution functions describing the state of plasma (i.e., Maxwell-Boltzmann velocity distribution, Boltzmann excitation formula, Saha ionization formula, Planck function) correspond to the same value of temperature. Such a state exists in the deep interior of the Sun and other stars. In other places in the universe the parameter in different distribution functions is different and one speaks about electron kinetic temperature, ion kinetic temperature, excitation temperature, ionization temperature, radiation temperature. In thermodynamic equilibrium all these temperatures are equal.

(c) *Energy Transfer*

Temperature decreases with the distance r from the solar center, more electrons are bound to nuclei, and opacity $\varkappa$ of the plasma increases. For distances $r > 0.88 \, R_\odot$ H becomes neutral. The sharp increase of opacity and of the temperature gradient (Equation (4.10)) caused by the neutral H is so large that the radiative transfer of the stored energy becomes inefficient and convective transfer sets in (Equation (4.11)).

In the construction of stellar models one has to compute the temperature gradient at every layer of the model for the radiative transfer (Equation (4.10)) and compare it with the adiabatic temperature gradient (Equation (4.11)) where dP/dr is given by Equation (4.7). As long as the adiabatic gradient is greater than the radiative gradient (in absolute values, both gradients are negative) the energy is transferred more efficiently by radiation. If, however the radiative gradient tends to become too large, i.e.

$$\left|\frac{dT}{dr}\right|_{\text{rad.}} > \left|\frac{dT}{dr}\right|_{\text{ad.}} \qquad \text{(Schwarzschild's criterium)} \qquad (4.14)$$

convection sets in which tends to reduce the radiative gradient to the adiabatic one.

(d) Convection, Chromosphere and Corona

The hot plasma clouds (about 1000 to 2000 km large) rising in the convective layer reach up to the lowest part of the solar atmosphere – to the photosphere (Figure 4.17). There they are observed for 8 min or so as granules (Figure 4.18), which radiate their energy, cool and sink downwards to the bottom of the convective layer. The rising plasma clouds in the convective layer contain energy in form of radiation, thermal motion and kinetic energy due to the rising motion – which is of about 0.4 km s^{-1}. Much of the energy content of the granules is radiated in the outer space as photons. The kinetic energy of the rising granules is transformed into acoustic noise in the photosphere.

The acoustic waves generated by granulation propagate outwards into the solar atmosphere to regions of lower density and are eventually transformed into shock waves. The shock waves easily dissipate so that their energy is transformed into the heat of the atmospheric plasma. The dissipation of shock waves explains why the temperature rises from about 4000 K at the top of the photosphere to 10^6 K in the corona. Thus the existence of chromosphere (Figure 4.19) and corona (Figure 4.20) is a consequence of the convective layer and its kinetic energy.

The energy from the chromosphere and the corona is emitted in form of electromagnetic waves ($\sim 10^{29}$ erg s^{-1}) and partly in form of the solar wind ($\sim 10^{27}$ erg s^{-1}). In brief, the atmosphere represents a transition between the storage of radiation and mass in the solar interior to a propagation regime in the interstellar space. It is a transition from nearly complete isotropy in the interior to complete anisotropy in the interstellar space.

(e) Magnetic Fields

The moving plasma in the convective zone is magnetized. Magnetic lines of force, frozen in the plasma, are stretched, twisted and compressed by convection and other types of plasma motions, such as turbulence, differential rotation etc. By the coupling of the field with solar plasma, the mechanical energy of motion is thus partly transformed into magnetic energy. Magnetic fields on the Sun are thus intensified and their intensification compensates for their dissipation.

The interaction between the magnetic fields and the solar plasma produces a variety of solar phenomena called solar activity: plages, sunspots, coronal condensations, prominences, flares, etc. Without the convective zone there would be no solar activity, no polar aurorae, no geomagnetic storms and ionospheric disturbances, so that many solar physicists and some geophysicists would be without a job.

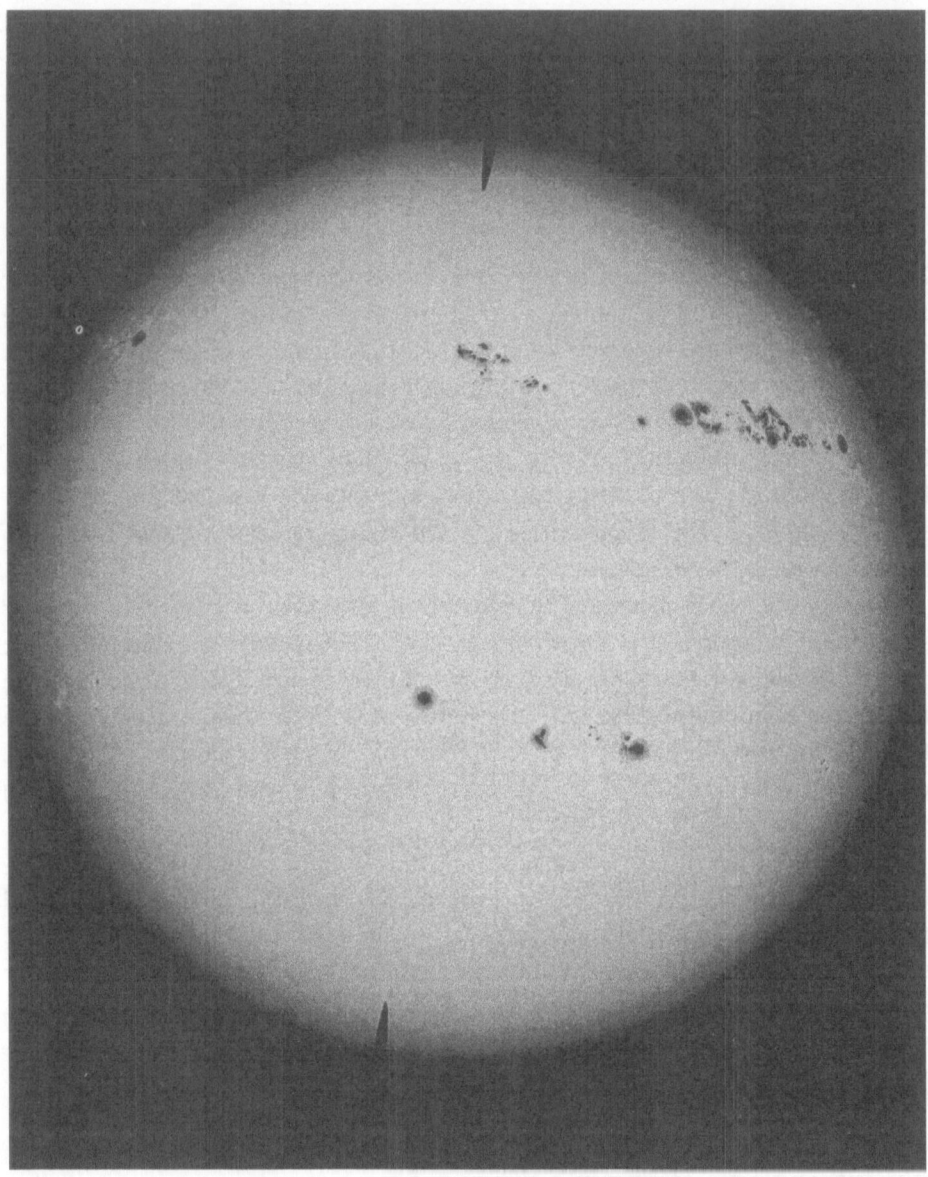

Fig. 4.17.   Photosphere is the lowest layer of the solar (and generally stellar) atmosphere. Though relatively thin, it radiates most of the solar luminosity. The photons are emitted by recombination of neutral H atoms with free electrons $(H + e^- \rightarrow H^-)$. Sunspots represent an electromagnetic interaction on a large scale: they are relatively cool regions with strong magnetic fields. (Mount Wilson and Palomar Observatory, February 20, 1956).

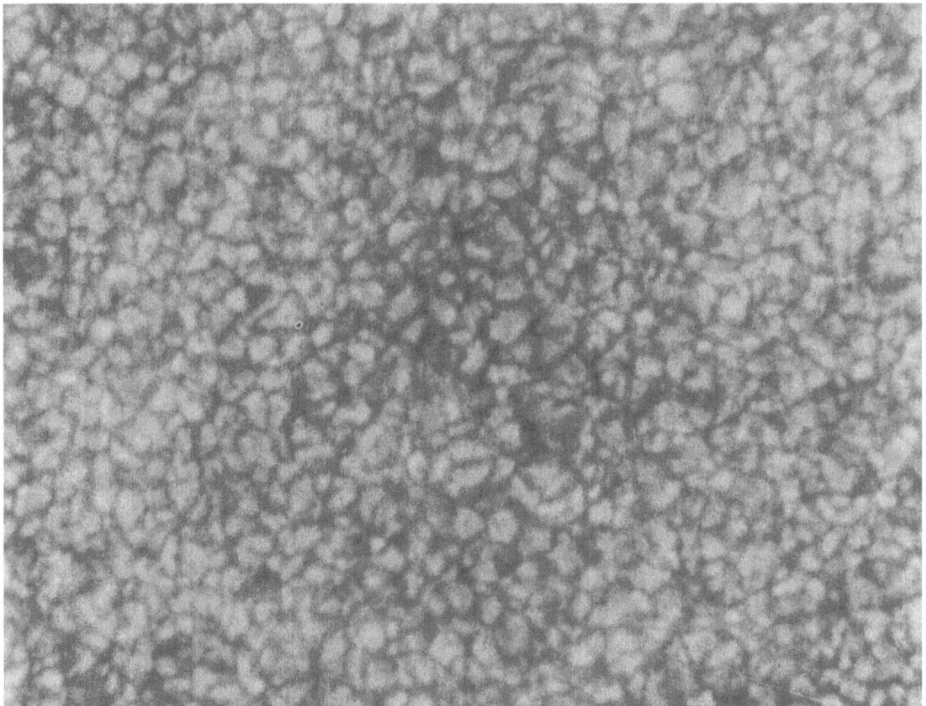

Fig. 4.18. Granulation. Convective currents transport hot plasma to the photosphere, where they are seen as granules. Their size is from 500 km to about 2000 km, they rise with a velocity of about 400 m s^{-1} and may be observed for about 8 min (Observatoire du Pic du Midi).

(f) *Stellar Convection and Magnetic Fields*

The study of stellar interior shows that the convective zone is more pronounced in the cooler stars and becomes weak for the stars with higher surface temperature, until it dies out at about 7500 K on the main sequence. In giants the convective zone exists for temperatures lower than 6500 K and in supergiants only when their surface temperature is lower than 5000 K.

Magnetic fields are associated with the convection in stars, as in the case of our Sun. Many stars of type G to M show emission lines Hα and II and K of ionized Ca; their emission is probably caused by plages connected with stellar magnetic fields like on our Sun. Other examples of stellar magnetic fields are found in flares on some cool main-sequence stars called flare stars. Starspots, an analogy to sunspots – are supposed to cause small luminosity variations in certain cool dwarfs (e.g. Ross 248). Variations of Hα in the spectrum of Betelgeuze has been explained by intense prominence activity in its atmosphere. These and other observations indicate that convection and magnetic fields are a common phenomenon on stars.

Gravitation is of basic importance for convection: the exchange of lower hot clouds with cooler upper plasma tends to minimize gravitational potential energy.

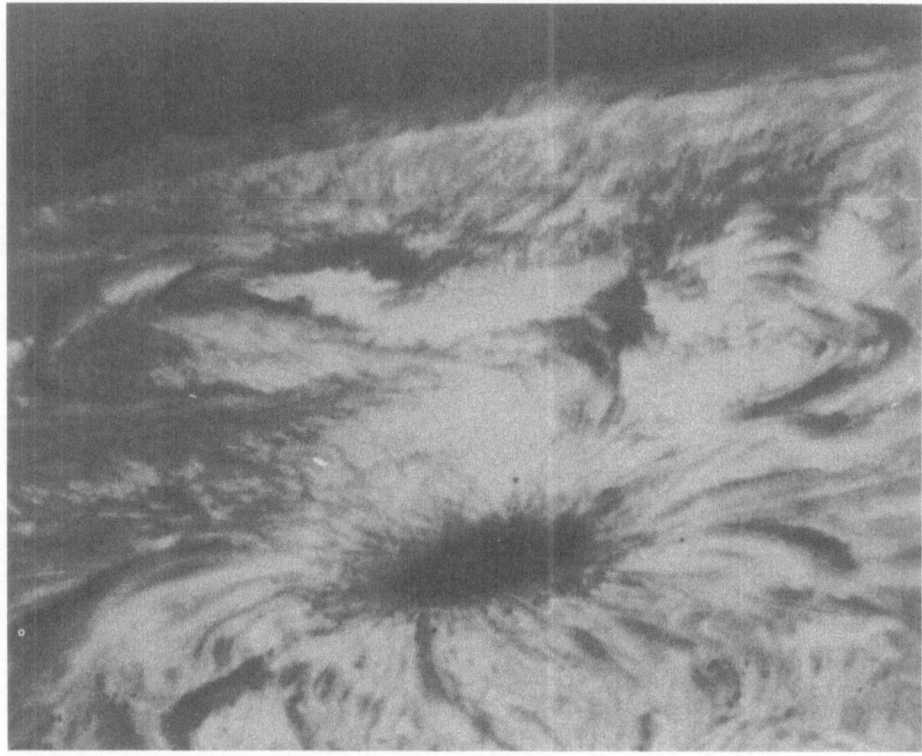

Fig. 4.19. Solar chromosphere as photographed in Hα line (Sacramento Peak Observatory). The photograph represents a small fraction near the solar limb. It is about 10 000 km high and its full height above the photosphere may be seen just at the limb. Chromosphere is inhomogeneous in density and temperature distribution. The regularity of some dark features is caused by strong magnetic fields of a spot.

4.5.2. WHITE DWARFS

White dwarfs are highly condensed objects of degenerate matter. Their internal structure is not dependent on temperature of the degenerate material. The self-gravitation of the star is balanced by pressure of degenerate electrons only.

(a) *Interior*

To determine the internal constitution of a white dwarf three equations are sufficient: the equation of hydrostatic equilibrium (4.7), equation of state for the degenerate gas (3.19) and Equation (4.9) relating mass M_r to density. There are three equations for three variables ϱ, P and M_r.

The integration proceeds outwards from the center, where one assumes a certain value for density ϱ_0. The interior of the white dwarf is divided into concentric shells of thickness Δr; the n-th shell has radius $r_n = n(\Delta r)$, pressure P_n, density ϱ_n and

Fig. 4.20. Solar corona is the highest and most extended layer of the solar atmosphere. Its hot plasma (about 10^6 K) is observed in X-rays, radio waves, in forbidden lines (Figure 3.4) or during solar eclipses as white-light corona; the latter is photospheric light scattered on coronal electrons. The photograph is from the 1954 June 30 eclipse (Astronomy Dept., Kiev University).

encloses mass M_n. The next shell has

$$\left. \begin{aligned} & r_{n+1} = r_n + \Delta r, \\[6pt] & P_{n+1} = P_n + \left(\frac{\mathrm{d}P}{\mathrm{d}r}\right)_n (\Delta r), \quad \text{where } \left(\frac{\mathrm{d}P}{\mathrm{d}r}\right)_n = -\frac{G\varrho_n M_n}{r_n^2} \\[6pt] & \varrho_{n+1} = \varrho(P_{n+1}) \quad \text{from equation of state,} \\[6pt] & M_{n+1} = M_n + 4\pi r_{n+1}^2 \varrho_n \Delta r. \end{aligned} \right\} \tag{4.15}$$

From the assumed central density ϱ_0 ($r_0 = 0$ and $M_0 = 0$) the equation of state gives central pressure P_0. Equations (4.15) give r_1, ϱ_1, and M_1. The numerical integration continues outwards to the point where pressure P is equal to zero. The corresponding values of r and M_r for this point represent the radius R and the Mass M of the white dwarf.

The integration is then repeated for different values of the central density ϱ_0, so that one receives one-parametric family of white-dwarf models with different mass and radius. It comes out that for an assumed central density there is a unique solution. On the contrary, for an assumed mass the solution need not be unique (see

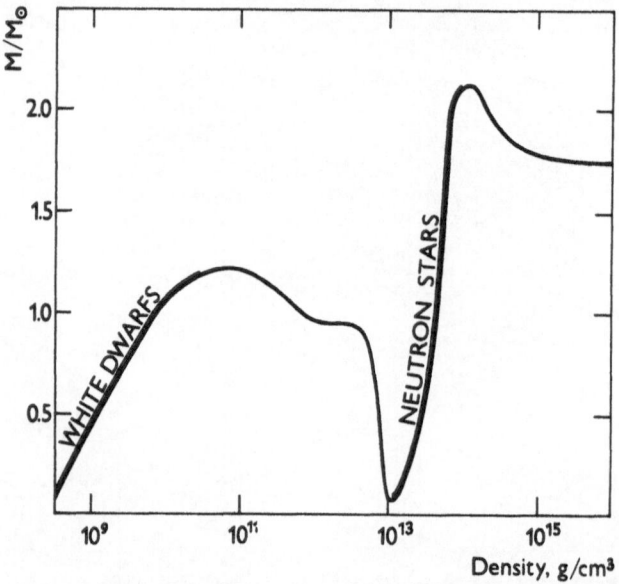

Fig. 4.21.   Mass of a degenerate star as a function of its central density. The points on negative slopes of the curve represent unstable configurations. Notice the limits for the stable configurations (white dwarfs and neutron stars). The same mass may be in form of a white dwarf or of a neutron star.

Figure 4.21). The larger the mass M of a white dwarf, the smaller its radius R (Table 4.2). The computed mass-radius relation can be compared with observations and thus the theory of the structure of white dwarfs can be tested.

The pressure inside white dwarfs is due to degenerate electrons and due to relativistic effects there is a limit to the mass than can be supported by such a pressure. Neutron degeneracy sets in when the limit of 1.4 $M_\odot$ is surpassed.

TABLE 4.2

Central density ϱ_c, mass M and radius R of white dwarfs

ϱ_c g cm^{-3}	M ($M_\odot = 1$)	$2R$ (km)
2.5×10^5	0.22	28 000
1.6×10^7	0.88	13 000
1.9×10^9	1.38	4 000

(b) *Atmosphere*

The degenerate matter of a white dwarf is blanketed by a plasma envelope. Though the plasma layer is thin (about 1% of the stellar radius), it is important for energy balance of the white dwarf and for astronomers too. It protects the hot degenerate interior against the chilly interstellar space. The interior conducts heat very well, so that it is practically isothermal. Without the opaque plasma envelope the white dwarf would cool very fast.

The visible part of the plasma envelope of a white dwarf, i.e., its atmosphere, emits radiation which brings information about temperature, chemical composition and surface gravity on the star. The surface gravity is about 10^3 to 10^4 times higher than in normal main-sequence stars, the atmospheric pressure is two orders of magnitude higher and therefore spectral lines are very broadened.

The structure of the non-degenerate layer may be computed as in the case of surface layers of normal stars. Integration of equations proceeds inwards, until the depth is reached where degeneracy commences to set in. There the condition in Equation (3.15) is fulfilled. The corresponding temperature in Equation (3.15) is practically the temperature of the isothermal interior of the white dwarf. It depends on mass, luminosity, chemical composition of the white dwarf and it is comparable with the temperatures in the Sun. The mass is also comparable which implies that the thermal energy content is also comparable. For the Sun it is 2.7×10^{48} erg which would be sufficient to sustain the present solar luminosity (2×10^{33} erg s^{-1}) for 10^{15} s or 3×10^7 yr.

(c) *Energy Sources of White Dwarfs*

The thermal energy of the interior of white dwarfs is the main source of their luminosity. More precisely thermal energy of the non-degenerate nuclei, because the degenerate electrons cannot give up any kinetic energy (due to Pauli principle).

Other sources of luminosity, such as gravitation or thermonuclear reactions, are of little importance or no importance at all. No gravitational energy is available: a white dwarf obeys the mass-radius relation, so that a further release of gravitational energy due to contraction is impossible. Thermonuclear H burning in the interior of a white dwarf ($\varrho \sim 10^6$ g cm^{-3}, $T \sim 10^7$ K) would generate energy at an enormous rate. The abundance of H in the interior of white dwarfs must, therefore, be very small ($< 10^{-4}$) to be consistent with their low luminosity. Thermonuclear energy production at the base of the non-degenerate envelope would produce pulsations of the star. A search for pulsations of white dwarfs has however so far been fruitless. Thus the nuclear reactions are unimportant and the stored thermal energy remains the only source of luminosity of white dwarfs.

The luminosity of white dwarfs is 10^2 to 10^3 times smaller than the solar luminosity. Their thermal energy should, therefore, be sufficient to supply the radiation of white dwarfs for periods of 10^9 or 10^{10} yr. White dwarfs thus emit the stored energy which the star accumulated during its previous stages of evolution. Their present evolution consists in slow cooling.

The pressure and densities in some normal plasma stars may be so high that the material of their central core is also degenerate. That is the case in red giants; a red giant is a white dwarf surrounded by an extended envelope. At the botton of the envelope nuclear fuel is burning.

4.5.3. Neutron stars

Neutron stars are built up of cold superdense matter. Their central densities are $\varrho_0 \gtrsim 10^{13}\ \mathrm{g\ cm^{-3}}$. The matter at such densities consists mainly of neutrons with a small admixture of electrons and protons. The neutron gas is degenerate. The size of neutron stars depends on their mass; it is about 10 km close to gravitational radius (Equation (2.67)).

(a) *Equations of Structure*

As in the case of white dwarfs, the structure of neutron stars is computed by numerical integration of equations for hydrostatic equilibrium and equation of state. High densities distort the space time so much that the Newtonian law cannot be used for the equation of hydrostatic equilibrium. General theory of relativity has to be used instead to express the strong gravitational force. Instead of Equation (4.7) the relativistic form of the equation for hydrostatic equilibrium is used, i.e.

$$-\frac{dP}{dr} = G\frac{\left(\varrho + \frac{P}{c^2}\right)\left(M_r + 4\pi r^3 \frac{P}{c^2}\right)}{r^2\left(1 - \frac{2GM_r}{c^2 r}\right)}. \tag{4.16}$$

Comparing the Newtonian equation with Equation (4.16), we may see that the relativistic gravitational forces are increased by the increase of density ϱ and mass M_r as well as by decrease of the denominator by the factor $(1 - 2GM_r/c^2 r)$.

The equation of state (Figure 3.11) relating density and pressure for degenerate neutron gas is not as well known as for degenerate electron gas in white dwarfs. Different theories give different results. Thus for the maximum mass of neutron stars values from 1.6 $M_\odot$ to 2.0 $M_\odot$ have been deduced, though 2.0 $M_\odot$ seems preferable. Rotation may cause a further increase of the limit mass of neutron stars. The collapsing stellar mass larger than 1.4 $M_\odot$ (i.e., maximum mass of white dwarfs) and smaller than about 2 $M_\odot$ should end as a neutron star (Figure 4.21).

(b) *Models of Structure*

According to the computed models the neutron stars are heterogeneous and consist of different layers (Figure 4.22). The density rises rapidly from the surface inwards, the electrons are degenerate and atoms completely ionized. Behavior of the upper thin layer ($\sim 10^3$ cm) is apparently strongly influenced by very intense magnetic fields.

Deeper down, at densities $> 10^5$ g cm^{-3}, the bare atomic nuclei arrange themselves into cubic crystalline lattice, as a consequence of their mutual repulsion in the densely packed material. At still higher densities $> 10^{11}$ g cm^{-3} (neutron drip point) neutrons evaporate from the nuclei. The number of free neutrons becomes larger than inside the nuclei. The solid lattice of neutron rich nuclei bound by electrostatic forces coexists with relativistically degenerate electrons and normally degenerate neutron gas.

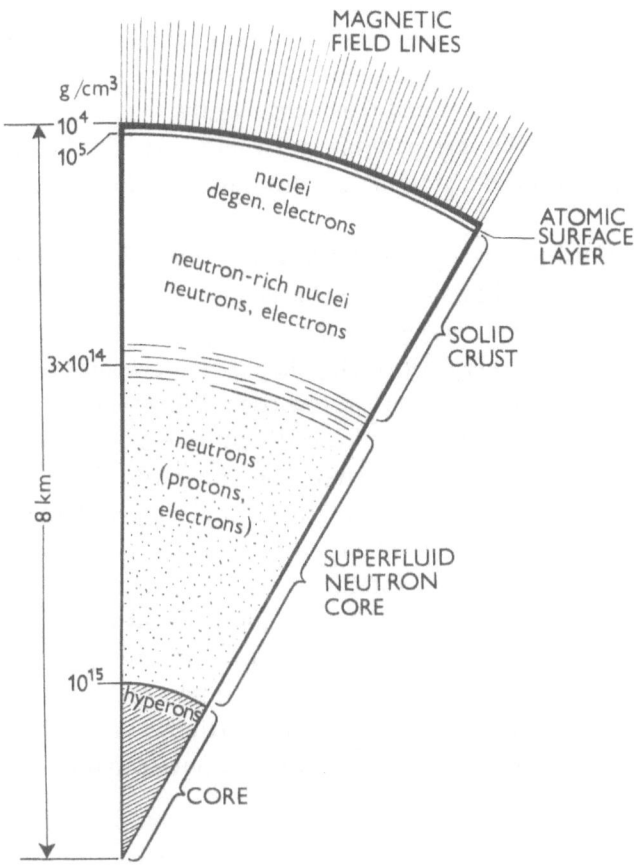

Fig. 4.22. Structure of a neutron star.

Still deeper – beneath the crust – at densities $> 10^{14}$ g cm^3 the nuclei dissolve completely and the matter consists of three different degenerate gases: most abundant non-relativistic neutrons, relativistic electrons and non-relativistic protons. At densities above 10^{15} g cm^{-3} the nucleons are partly excited to hyperon states (hyperon gas) and electrons to muons. The Figure 4.22 represents very schematically a cross section of a neutron star, but the reader is asked to view it still with a healthy distrust.

(c) *Pulsars*

At present it is generally accepted that pulsars are neutron stars with extremely strong magnetic fields (10^{10} to 10^{12} G). The neutron stars originate by collapse of middle weight stars at the end of their thermonuclear evolution (Section 5.4). The angular momentum and the total magnetic flux are conserved during the collapse. During the process distances of all particles from the rotation axis of the collapsing star decrease substantially (roughly 10^4 times) which means that the angular velocity ω has to increase $\sim 10^8$ times if the angular momentum ($\sim mr^2\omega$) is to be conserved. This would explain the very short rotation period of pulsars.

The magnetic field is frozen in the plasma of the parent star, so that the total flux is conserved during the collapse. The contraction from the radius 10^6 km to 10 km decreases the surface to 10^{-10} and thus enhances the field about 10^{10} times. The surface fields of normal stars are in the interval from 1 to 10^4 G. The extremely strong magnetic fields of neutron stars (pulsars) are thus inherited by collapse from the parent stars.

The flashes of light from a pulsar are probably emitted by bunches of charged particles corotating with the pulsar magnetosphere (Figure 4.23). If far enough from the rotation axis and close enough to the light cylinder – where the velocity of corotation $\omega r = c$ – the charged particles should emit synchrotron radiation. The radiation is strongly directional in a narrow cone. Whenever the conical emission beam is directed towards us, flashes are recorded by terrestrial radio telescopes. Thus only a fraction of pulsars may be observed from the Earth, viz. those with rotation axis normal to the line of sight or nearly so. Neutron stars in close binaries may be observed as X-ray pulsars (Figure (4.24).

Other models of pulsars, their internal structure, magnetosphere and emission mechanisms have been proposed. They will not be discussed here, because there seems to exist no reason to prefer a particular one. Moreover many theories of one phenomenon may lead to confusion in minds of non-specialists. There are still questions to be answered but it is generally accepted that pulsars are neutron stars and that they are genetically connected with supernova explosions and supernova remnants.

(d) *Properties of Matter in High Magnetic Fields*

The magnetic fields on the surface of pulsars are several orders of magnitude stronger than the strongest fields reached for a short time in terrestrial laboratories, which is about 10^7 G.

Strong magnetic fields influence the structure of atoms and molecules. A substantial change should for example occur in a H atom placed in a field B_0 such that the Larmor radius (Equation (3.5)) for an electron is comparable with the Bohr radius $a_0 = 0.5 \times 10^{-8}$ cm of the electron orbit. Then the Lorentz force becomes more important than the electrostatic interaction between the electron and the proton. This occurs for fields $B > B_0 = 2 \times 10^9$ G. As a result, the normally spherically

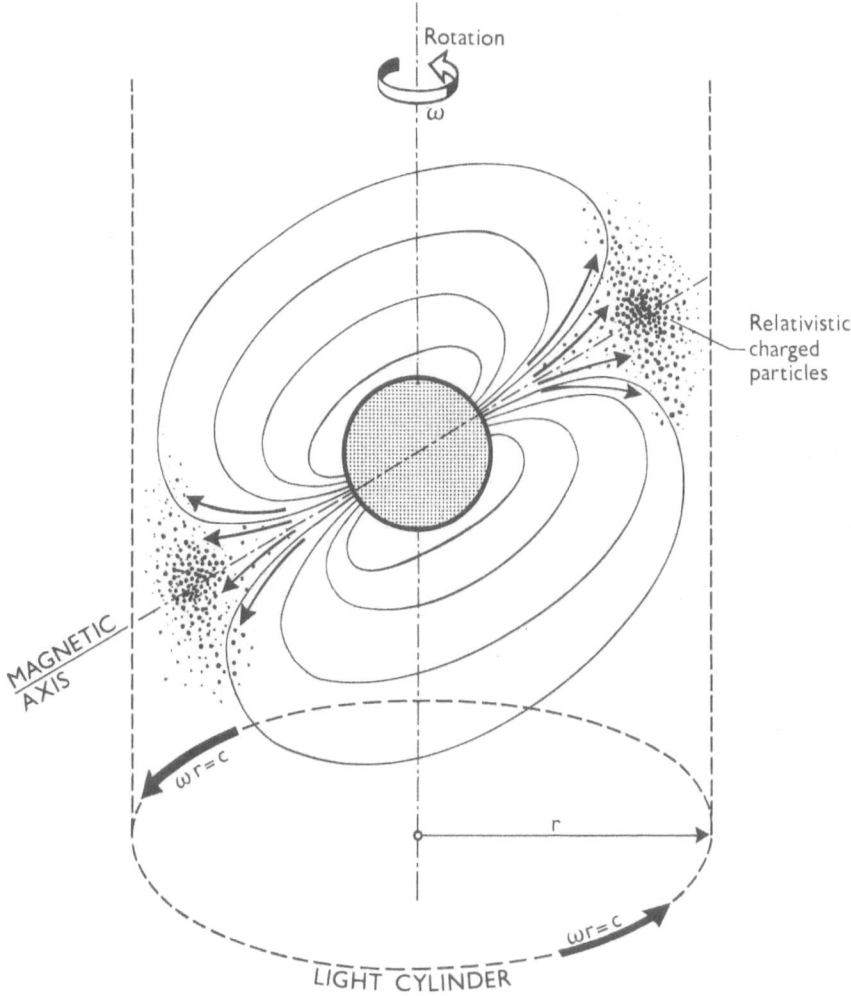

Fig. 4.23. Pulsar is a fast rotating neutron star endowed with a very strong magnetosphere. Particles at the periphery of the magnetosphere have relativistic velocity and emit synchrotron radiation.

symmetric atoms become elongated in the direction of magnetic field lines. Their length and the binding energy of electrons increase with the intensity of the magnetic field.

Nuclei of the elongated atoms (Figure 4.25) are not as well shielded by their electron envelope, as in the case of spherical normal atoms. Molecules with high binding energy may be formed. In the presence of very strong magnetic fields polymer chains will be formed, oriented along the lines of force. The polymer chains cannot be turned around an axis perpendicular to the magnetic field. From these examples it may be seen that physical and chemical properties of matter will be substantially changed by presence of strong magnetic fields. Such changed properties influence the behavior of matter on pulsars.

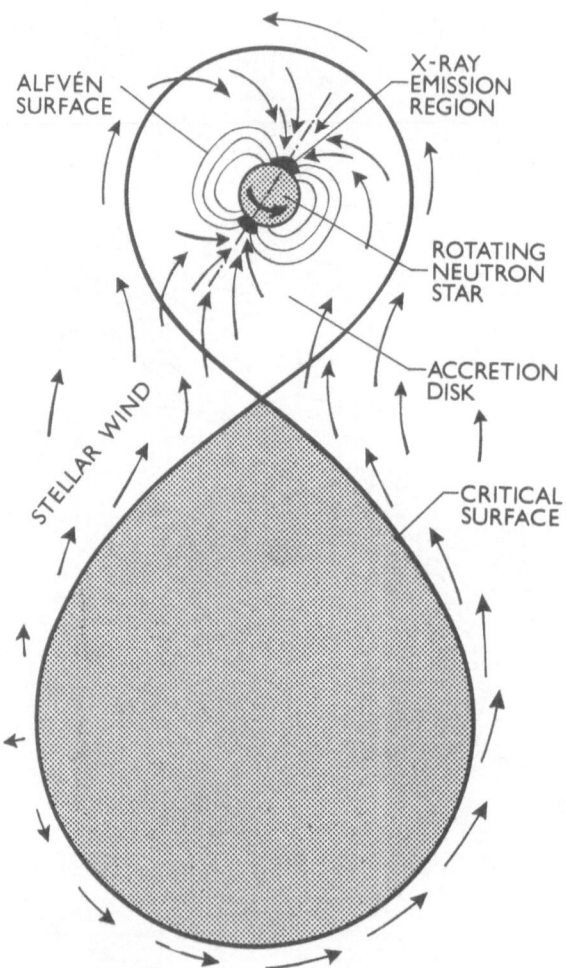

Fig. 4.24. Pulsating X-ray star. A neutron star in a close binary may be a source of X-rays. The plasma from the normal star is transferred and falls towards the small neutron star. It passes first through the critical surface (marked by a heavy curve – in the form of stellar wind or as an expanding envelope of the normal star). By falling on the neutron star the potential energy of the plasma is changed into heat and radiates eventually in X-rays.

4.5.4. BLACK HOLES

In discussing objects of stellar mass, one should mention black holes. Collapsed bodies with a mass larger than about 2 $M_\odot$ should be smaller than their gravitational radius; it implies that neither photons nor particles can escape from their surface – therefore the name 'black holes'.

A black hole probably exists in the binary star known as Cyg X-1 source (i.e. optical HDE 266 868). It coincides also with a weak variable radio source. This binary star contains a normal B0 supergiant and an invisible companion. The companion is at least as massive as the supergiant (masses of the B0 I supergiants are ~50 $M_\odot$). If it is invisible, it violates the mass-luminosity relation for normal stars. It

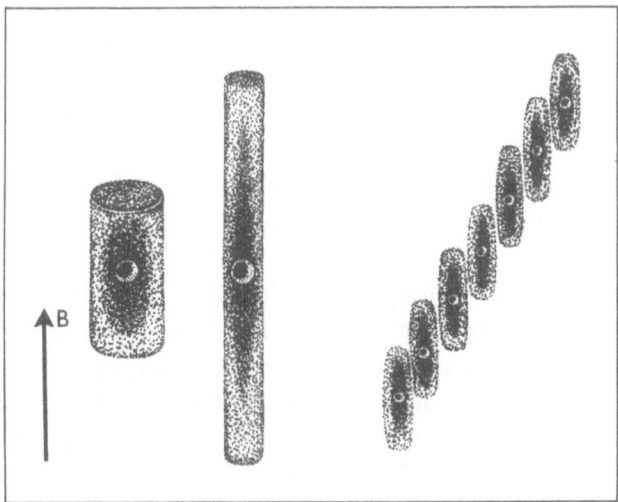

Fig. 4.25. Atom of H in a field of 10^{12} and 10^{13} G. Elongated atoms form long polymer chains with anisotropic properties. One cannot rotate the chain around the axis perpendicular to the magnetic field without breaking it (Kadomtsev).

turns out that the Cyg X-1 source may be a black hole accreting plasma from the supergiant. The potential energy of the accreted plasma is changed into heat and the hot streams rushing towards the black hole emit X-ray photons.

The properties of matter in a black hole are not known and at present nothing can be said about its structure. During the collapse all particles of the pre-collapse star disappear. So far as one can tell, the black holes are objects characterized exclusively by mass, angular momentum, electric charge and by nothing else. Four classes of black holes might therefore exist: Schwarzschild black hole with mass only, without electric charge and non-rotating; Kerr black hole which has both mass and angular momentum; non-rotating black hole with mass and electric charge; the general class with mass, angular momentum and electric charge. Transition from one class to another is possible, (e.g. by accretion of particles from outside).

4.6. Star Systems

Stars are bound by gravitation to different types of stellar systems. The systems of stars represent the higher structural level, at which individual stars themselves are constituent units. The number of stars in a system may be different: from two in binary systems up to 10^{12} in giant galaxies. The term 'star system' is thus a very broad one and includes multiple stars, star clusters and galaxies.

Gravitation is the binding force of all the stellar systems. Mutual gravitational attraction binds the system members into one unit. The gravitational potential is an important characteristic of the system as it is a measure of its stability. If the kinetic energy of its members is larger than the total potential energy, the system will disperse soon after it has been formed (as is the case in stellar associations). If, on the other hand, the gravitational attraction exceeds the dispersive tendency of the member motions, the system is more stable and more lasting (as is the case in globular clusters).

Smaller stellar systems are members of larger systems and galaxies themselves are members of clusters of galaxies, which seem to be the largest structural units in the observable universe.

The grouping of stars is a consequence of their formation: stars – at least most of them – are formed in groups. Single stars are less frequent than double or multiple stars. Thus in our nearest neighborhood there are five binary or multiple systems, one planetary system (that of our Sun) and only one simple star. But even a simple star may have an unseen companion which is difficult to detect. Statistics indicate that for every 30 single stars there are at least 47 double stars with 94 components and 23 multiple stars with 81 components. Thus from 205 stars only 30 are single stars and all the others constitute double or multiple systems.

(a) *Associations*

Loose groups of hot stars (type O and B) or of variables of the type T Tauri and RW Aurigae are called stellar associations. An association consists of a few dozens (or even hundreds) of stars dispersed in a large volume among other stars. The associations are very unstable (due to low density) and, therefore, very young groups of stars of common origin. Their age, determined from velocities of expansion, is in some cases $\sim 10^6$ yr. After about 10^7 yr they disappear completely. Associations prove that stars are formed in groups and that stars are formed at present times.

About 50 OB associations and 25 T associations have been identified on the sky. Their total number in our Galaxy will, of course, be much larger. Associations have been also discovered in other galaxies.

(b) *Open Clusters*

Agglomeration of stars in space, the average density of which exceeds the density of background stars, is called a star cluster. By density of stars in this context one understands the number of stars in a unit volume, e.g. in one cubic parsec. There are two types of star clusters: open clusters (galactic clusters) and globular clusters which contain substantially more stars.

An open cluster is a more compact group of stars than associations are. But the number of its members ($\sim 10^2$ to 10^3) is not sufficient to secure its stability. The density of stars (i.e. number in a cubic parsec) in open clusters is substantially higher than in the solar neighborhood. Open clusters are found near the galactic plane (therefore they are called galactic clusters) and if young in or near spiral arms. Ten hundreds of open clusters are known in our Galaxy and their total number may be several thousands. Open clusters are also observed in other galaxies.

Thé stars of the same cluster are approximately of the same age. But hot stars with high luminosities (upper left part of the main sequence) burn out their H much faster and, therefore, leave the main sequence sooner than the less luminous stars. The turn-off point on the main sequence of a cluster hence indicates its age. A star leaves the main sequence when it has transformed H to He in 12% of the star's mass. Its age is then

$$T \approx 10^{10} M/L \text{ yr} \tag{4.17}$$

where M and L are mass and luminosity of the star expressed in solar units. A majority of open clusters have ages from 10^7 to 10^9 yr.

(c) *Globular Clusters*

Globular clusters contain a considerably large number of stars than open clusters – generally from 10^5 to 10^7 members. They are a compact system with a large (negative) potential energy and are therefore very stable. By close encounters between stars in the globular cluster some may gain escape velocity; they then leave the gravitational field of the cluster, and take away some of its kinetic energy. As a result the cluster becomes still more compact and more stable. About 125 globular clusters are known in our Galaxy, and their total number may be about five hundred. In contrast with the galactic clusters they are found far from the galactic plane, forming a spherical subsystem of our Galaxy with concentration towards its center.

The branch of giants turns off from the main sequence at the locus of the F stars – which corresponds to ages of about 10^{10} yr. The globular clusters are much older than the galactic clusters. It is possible that condensation of Galaxy and of all the globular clusters occurred simultaneously, about 10^{10} yr ago. Globular clusters are very old, containing the oldest stars that lived through the whole history of our Galaxy.

4.7. Galaxies

Galaxies are large and massive systems of many different objects held together by their mutual gravitational attraction. A galaxy contains stars, double and multiple stars, star clusters, nebulae, molecular clouds, interstellar dust, high-energy cosmic rays and extended interstellar magnetic fields. Vast interstellar spaces of a galaxy contain photons of all energies – from γ photons to radio waves. They are mainly emitted by the constituents of the galaxy itself, i.e. by its stars, nebulae, interstellar matter and cosmic rays. If not absorbed in the galaxy they escape into the intergalactic space.

The structure of galaxies is usually characterized by total mass, population, luminosity, radius, morphological type, ellipticity, total angular momentum and kinetic energy. In nearby galaxies the detail structure may be studied by radio and optical techniques. Most of our knowledge of the detailed structure of galaxies comes from the Local Group of Galaxies, the most distant of which is about 1 Mpc from us.

The best known galaxy of the Local Group is the Andromeda Galaxy M31, a giant spiral (Figure 4.27). Another giant spiral is our own Galaxy (Figure 4.26). There are about 15 dwarf elliptical galaxies and four irregular galaxies in the group.

The Hubble classification of morphological types classifies about 95% of galaxies. The sequence of types corresponds to the increasing degree of flattening: elliptical (E_0 – E_7), lenticular (SO), spiral (Sa, Sb, Sc), barred spiral (SBa, SBb, SBc with increasing openness of spirals) and irregular galaxies (Ir I, Ir II of populations I and II). The sequence reflects the initial conditions of galaxy formation: specific angular momentum increases along the sequence. Rotation plays an important role in formation of galaxies (Section 5.3).

Other classifications of galaxies have been later proposed and new important types have been added, such as Seyfert galaxies, N-type galaxies, compact galaxies, quasistellar galaxies, radio galaxies and quasistellar sources. Their genetic relations are not yet completely understood. The origin of individual types of galaxies remains an open question for astrophysics and cosmology.

The important structural features of galaxies are: stellar population, spiral arms, nuclei and their activity.

(a) *Star Populations*

The stars, gas and dust of which galaxies consist show a spatial arrangement into several subsystems or population types. The populations differ in total mass, flattening, age and chemical composition (Table 4.3). Extreme population I consists of very young objects such as supergiants, diffuse nebulae, reflection nebulae, associations and youngest open clusters. The members of the extreme population I are distributed in a very thin layer in the galactic plane and they move in almost circular orbits around the galactic center. The thickness of the population space is about 300 pc. The members of this flat population revolve about the center of our Galaxy with a velocity of about 250 km s^{-1} (in the neighborhood of the Sun), which is five times faster than for the nearly spherical halo population II.

The members of the halo population II have orbits strongly inclined toward the galactic plane and exhibit high velocities perpendicular to that plane. Globular clusters, subdwarfs and RR Lyr variables (with period >0.4 days) are the main components of the halo population. They represent the archaic component of our galactic system. Between the flat population I and the halo population II there are three intermediate population types, as seen from Table 4.3. They were formed during different stages of galactic evolution.

(b) *Spiral Arms*

The majority of non-elliptical galaxies has well defined spiral arms. The different objects which define the spiral arms belong to the flat population I and they are called 'spiral tracers': supergiants, associations, open clusters, neutral H observed at λ 21 cm, diffuse nebulae, dust lanes and some others. The structure and position defined by the different spiral tracers are not necessarily the same.

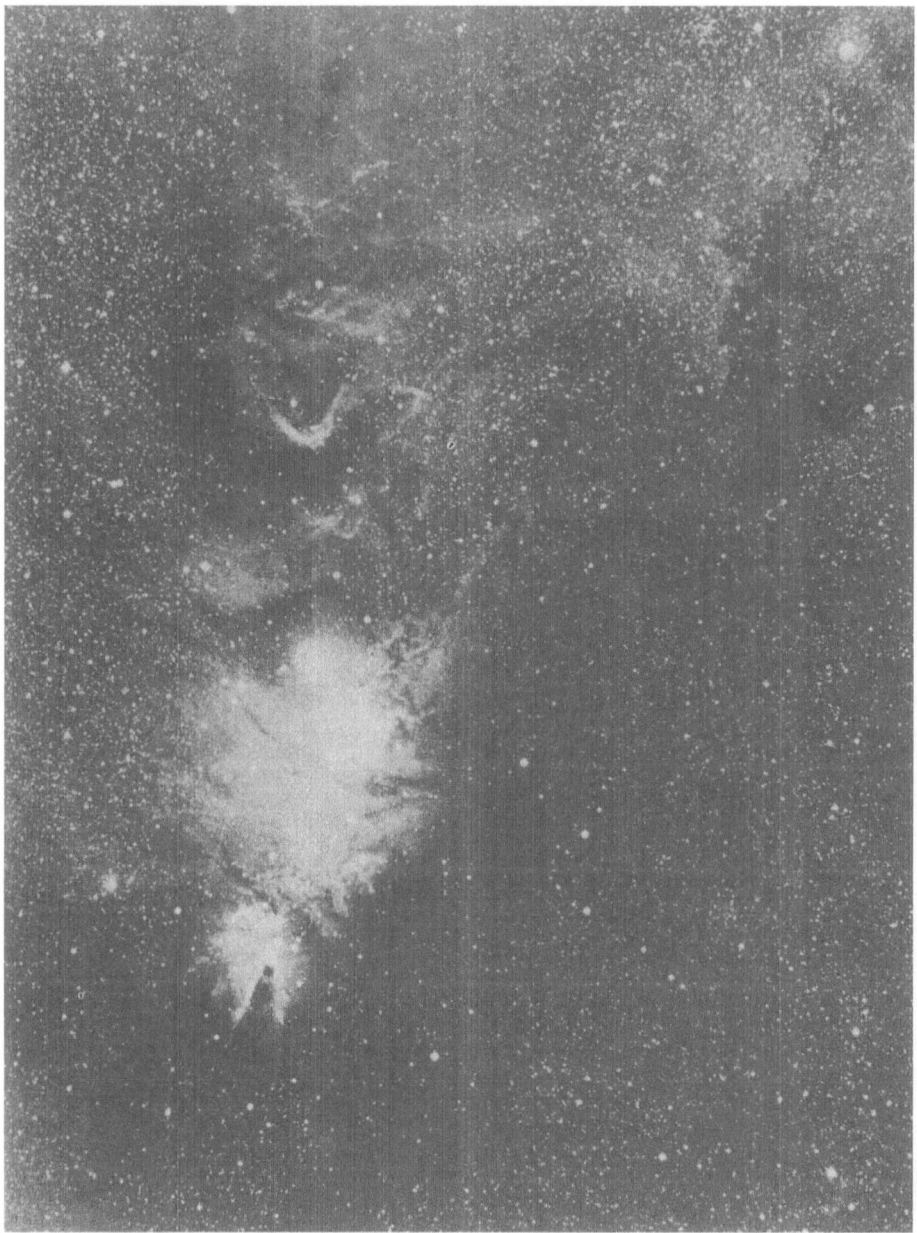

Fig. 4.26.   Our Galaxy seen from inside (a section of Milky Way). Besides many stars which represent the main component of the Galaxy, one observes emission nebulae (such as NGC 2264 in this photograph) i.e. luminous interstellar gas illuminated by nearby hot stars. Also dark patches of cold interstellar dust can be seen (Karl Schwarzschild Observatorium, Tautenburg).

Fig. 4.27.   M 31 – Andromeda Galaxy. Though about 2 millions of light years distant, it may still be seen by the naked eye. It is larger than our Galaxy and is the largest member of the Local Group of galaxies. In 21 cm-radiation it appears considerably larger than in light, since it is surrounded by large amounts of neutral H. To the west of M 31 is a small elliptical galaxy NGC 205 (Karl Schwarzschild Observatorium, Tautenburg).

TABLE 4.3

Population types and their five subdivisions. Approximate values of total mass, of the age, ratio of heavy elements to H, flattening (axial ratio) and some examples of objects are given for individual subdivisions

Population	Mass $10^9 M_\odot$	Age 10^9 yr	Heavy el. hydrogen	Axial ratio	Objects
I – Extreme	5	<0.1	0.04	100	Interstellar gas Associations T Tauri stars Supergiants
I – Older	10	0.1–1.5	0.02	50	Strong line stars A stars Older open cluster
II – Disk	40	1.5–5	0.01	20	Planetary nebulae Novae Weak-line stars
II – Intermediate	40	5–8	0.004	5	High velocity stars Long period variables
II – Halo	20	>8	0.001	2	Globular clusters Subdwarfs RR Lyrae variables

About a half of the observed galaxies show more than two arms. Multiple arms prevail in the outer parts of spiral galaxies while close to the nucleus two distinct arms are normally seen. Sometimes it is difficult to discern the spiral pattern of a galaxy, because of the bifurcations of arms, dark streaks, links between arms, and secondary branches. A great variety exists in spiral structures and it may be sometimes difficult to assign a type to an observed galaxy.

If a spiral arm were a persistent feature of a galaxy, held together by magnetic fields, it would wind up into progressively tighter spirals. An additional turn would be added in about 10^8 yr as a result of galactic rotation. A persistent spiral arm would have a few dozens of turns – which contradicts the observations ('persistence dilemma'). The arms are ephemeral structures which have to be continuously revived. Today it is generally accepted that the spiral arms represent the crest of a density wave propagating through the galactic disk (Figure 4.28). The spiral structure is caused by a spiral gravitational field; the magnetic fields $\sim 3 \times 10^{-6}$ G are too weak to hold the spiral arms together. The density wave in the axially-symmetric disk is formed when the material (stars and especially interstellar medium) passes through the spiral potential trough.

The spiral potential waves move in the direction of galactic rotation, but more slowly than the material of the disk does. According to this spiral density wave theory the stars and interstellar matter are not attached to the arms, but rather they move from one arm to another in about 2×10^8 yr (at the Sun's distance from the galactic center). Once the spiral arms have been established, they can survive for 10^9 yr, i.e.

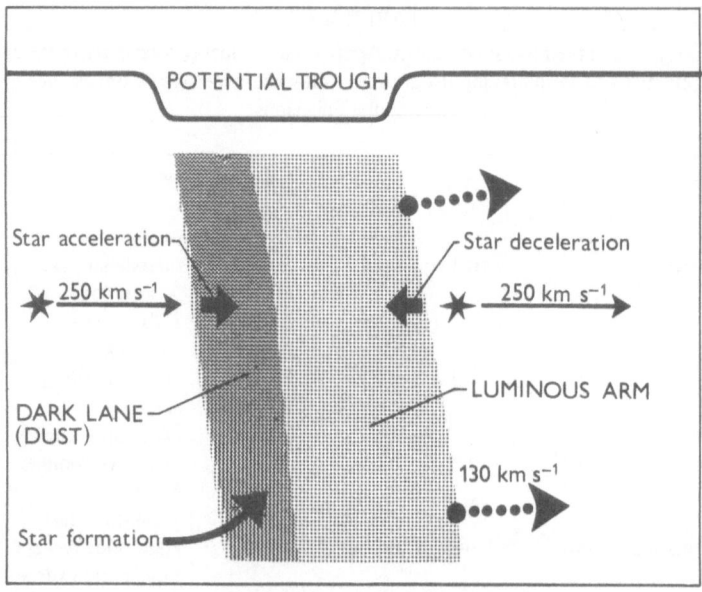

Fig. 4.28.   Section of a spiral potential wave (gravitational density wave). A star passing through the potential trough is shown.

several revolutions of the Galaxy (at the Sun's distance). However, there is not yet a satisfactory explanation how the spiral pattern arises.

The theory of gravitational density waves just described is not the only theory explaining the spiral structures. Some astronomers argue that spiral arms represent the tracks of material ejected from nuclei of galaxies, while others are of the opinion that, after all, the magnetic fields do play an important role in the spiral structure of galaxies.

(c) Galactic Nuclei

A small region of high luminosity is observed in the center of many galaxies. It appears as a starlike or nearly starlike bright spike which often coincides with the center of symmetry of elliptical or spiral galaxies, with the maximum star concentration and with the center of rotation of the galaxies. The region – called galactic nucleus – plays an important role in evolution of galaxies, though relatively very small. For example the Seyfert galaxy NGC 4151 (photographed by a balloon-borne telescope) shows a nucleus smaller than 7pc and a luminosity of 2×10^{43} erg s^{-1}, comparable with the total luminosity of our whole Galaxy. The nucleus of M 31 in Andromeda is elongated with dimensions 5×9 pc.

The position of the *nucleus of our Galaxy* could be determined within a few seconds of arc. It coincides with a radio source Sgr A West. Only an upper limit can be estimated for the mass of the nucleus, namely $10^8 M_\odot$. Nothing is known at present about the nature and the mechanism which enables the nucleus to expel vast expanding masses of gas: (1) The expanding arm at a distance 3 kpc from the center moves with velocities -53 km s^{-1}. It is between us and the nucleus and contains about 10^7 $M_\odot$ of neutral H. (2) Another arm expanding with velocities $+135$ km s^{-1} contains about the same amount of neutral H. Both phenomena indicate that gaseous masses of the order of 10^7 $M_\odot$ have been thrown out asymmetrically from the galactic nucleus some 10 to 15×10^6 yr ago. (3) High velocity clouds are found outside the galactic plane with mass $\sim 10^5$ $M_\odot$ and 1 to 3 kpc away from the nucleus. (4) Dense molecular clouds at distance 100 to 700 pc from the nucleus with radial motion about 200 km s^{-1} represent a mass of 10^6 $M_\odot$. The molecular clouds have dimensions of a few parsec and densities 10^3 to 10^5 H$_2$ molecules per cm^3. Their velocities and distances imply that they have been ejected from the nucleus about 10^6 to 2×10^6 yr ago. (5) Absorption features exist behind the nucleus with radial velocities of $+50$ km s^{-1} and with a mass of 10^5 $M_\odot$. The broad absorption in OH and H$_2$CO may be caused by gas ejected from the nucleus and seen against the source Sgr A East, which itself is only some 5pc behind the nucleus. The gas should have been expelled very recently.

(d) Activity of Galactic Nuclei

The phenomena observed in our Galaxy thus indicate that the nucleus of our Galaxy has been active for (at least) the last 15×10^6 yr. Our Galaxy belongs to galaxies with relatively moderate outflow of gases from their nuclei. Other well studied examples are M 51, M 31, NGC 253.

More violent activity is observed in the nuclei of Seyfert galaxies. Their very luminous nuclei with high excitation emission-line spectra show velocities in the range 500 to 4000 km s^{-1} or even more.

Closely related to the Seyfert galaxies are the N galaxies with a brilliant starlike nucleus in which most of the luminosity of the galaxy is concentrated. The nucleus is embedded into a faint nebulous envelope. Many N galaxies are strong radio sources and some are variable in light. Another type of extragalactic objects probably related to the nuclei of Seyfert galaxies and N galaxies are *quasistellar objects*, i.e., quasistellar galaxies (QSG) and quasars. They are starlike objects with very large redshifts in their spectra. Their luminosity is much higher than that of Seyfert galaxies. However the observed properties of all the three types of extragalactic objects (spectra, colours, variability) are so similar that their structure and energy generation are likely to be the same. The difference would be in scale only. Nuclei of Seyfert galaxies and N galaxies are probably small quasars and all quasars are apparently highly excited nuclei of giant galaxies. In quasars (and quasistellar objects in general) the extreme of violent nuclear activity is observed.

There is a large evidence that many other galaxies eject large amounts of matter out from their nuclei. For example the galaxy M 82 has a nucleus with violent activity (Figure 4.29). It is also a radio source designated in the Third Cambridge Catalogue as 3C 231. About 5×10^6 $M_\odot$ of gas (mostly H) have been ejected outward with velocities of about 1000 km s^{-1}. The velocity of the ejected gases increases with their

Fig. 4.29.   Irregular galaxy M 82 in Ursa Major constellation. It is about 10 million light years distant. There is much dust across the face of M 82. This galaxy is a radio source. Large-scale explosive activity takes place in its nucleus.

distance from the nucleus, reaching 2700 km s^{-1} at a distance of 4 kpc. One may deduce that the outburst of the nucleus must have started about 1.5×10^6 yr ago. The galaxy is filled with gas and the nucleus itself is not well seen in light. Nevertheless, it has been found in the IR; it coincides with the center of explosion and also with the center of magnetic fields determined from polarization vectors.

The immense radio fluxes from some extragalactic radio sources (radio galaxies and quasars) are a consequence of violent activity of the nuclei of the parent galaxies. Their radio luminosity may be as high as $10^{47} \text{ erg s}^{-1}$. The lifetime for the relativistic electrons is estimated to be about 10^6 yr (10^{13}s), so that the total energy radiated by a strong radio source in radio waves is $\sim 10^{60}$ erg. The efficiency of transformation of the liberated energy into the radio waves is roughly estimated to be about 0.01. Thus the energy necessary to account for the most powerful extragalactic radio sources may be $\sim 10^{62}$ erg. That is equivalent to conversion to He of a mass of H $\sim 10^{10} \, M_\odot$. The thermonuclear reactions have a relatively small efficiency (10^{-3}) of generating energy out of the rest energy of matter. Gravitation is by two orders of magnitude more efficient than thermonuclear reactions, and a more promising source for explosions of galactic nuclei (efficiency up to $\frac{1}{2}$ or so). Another candidate and the most efficient source of all might be annihilation of matter and antimatter (efficiency 100%). Whatever is the nature of the energy source in galactic nuclei, there is observational evidence that the nuclei of many (and probably of all) galaxies are dense, liberate large amounts of energy, generate relativistic particles and eject large amounts of matter – sometimes in the form of coherent objects.

What is the nature of galactic nuclei and quasars? What is the ultimate source of the immense quantities of energy liberated? There is no lack of hypotheses (e.g., supermassive star or superstar with mass 10^5 to $10^{10} M_\odot$ producing luminosity 10^{45} erg s^{-1} by thermonuclear reactions; supermassive black hole formed by gravitational collapse of a dense star cluster; frequent collisions of stars at high velocities and their disruption or coalescence to a massive body – which later explodes as supernova; formation of galaxies proceeds 'from inside out', by explosion of the nuclei or 'lagging cores from the early Big-Bang', 'white holes' or 'embryos of galaxies').

According to observations the major constituents of galactic nuclei are gas, plasma, dust, stars, electromagnetic radiation. Until now there is no observational evidence for or against other forms of matter – which simply means that the structure of galactic nuclei is difficult to observe.

4.8. Clustering of Galaxies

In the hierarchy of the universe, galaxies themselves are units in higher systems. Isolated galaxies are rare. Common are multiple galaxies and groups of galaxies with a small number of members (of the order 10^1) and size 1 Mpc to 3 Mpc. Within 16 Mpc in our neighborhood there are 55 groups of galaxies, but only very few isolated galaxies which could not be assigned to a definite group. Some groups – mainly those containing elliptical galaxies – are imbedded in a luminous halo with size up to 1.3 Mpc. The absorption lines in distant quasars may be caused by absorption in the large halos of the intervening groups of galaxies.

The Local Group contains about 21 thus far identified objects, among them our Galaxy and the Andromeda Nebula M31. Diameter of the Local Group of galaxies is about 1 Mpc.

Clusters of galaxies are larger and more massive systems than groups of galaxies (Figure 4.30). Their diameters range from 2 to 5 Mpc. The number of members may be low ($\sim 10^2$) for poor clusters of galaxies. In a rich cluster the number of galaxies may be 10^3. Some clusters are regular, with the number of galaxies (per unit volume) increasing towards the cluster center. Elliptical galaxies prevail in the regular clusters. On the other hand, an irregular cluster shows no marked concentration of galaxies and spirals and irregular galaxies prevail.

Clusters of galaxies are often considered to be the basic units in the large scale structure of the universe. They are easy to recognize at large distances and are therefore used to construct the velocity-distance relation. This relation found from observations represents a tangible proof that the universe is expanding.

Fig. 4.30.   Coma cluster. This cluster of galaxies contains about eight hundred galaxies – many of which may be seen in this photograph. The cluster is about 350 million light years distant from us and it recedes with a velocity of $6700 \, \mathrm{km \, s^{-1}}$ from us. The brightest member of the cluster is an elliptical galaxy (NGC 4889) of magnitude 12, near the center (Karl Schwarzschild Observatorium, Tautenburg).

The largest known systems of galaxies have $\sim 10^5$ members, and size from 30 to 60 Mpc, depending on the distance scale adopted. Such systems are called clusters of clusters, superclusters or cloud of galaxies. The Local Supercluster is of this size. Its center lies in or near the Virgo cluster. A total mass of 10^{15} $M_\odot$ has been derived from the rotation of the Supercluster. Our Galaxy is on the outskirts of the Local Supercluster, i.e. near its southern edge.

Many groups and clusters of galaxies are unstable systems. The relative velocities of galaxies in the cluster or group (and hence their total kinetic energy) are often too high. The observed masses (and the deduced potential energy) are then too small for the cluster or group to be held together by gravitation. Are there some forms of invisible (e.g., intergalactic) matter in the clusters to increase its potential energy and hence their stability? Are the velocities of galaxies in the clusters high from the beginning? The decay times of clusters of galaxies are $\leqslant 10^9$ yr.

Conspicuous instability must exist in chains of galaxies, i.e. in groups where the galaxies are arranged one after another. If a member of the chain has velocities of many thousands of km s^{-1} with respect to the other members, it must escape from the chain. Such a fast galaxy had to be ejected from a nucleus of some very active member of the chain.

4.9. Large Scale Structure of the Universe

There is a saying among the people of Indochina 'The world reaches only as far as a man can walk.' Its paraphrase into the astronomers' language would be 'The universe reaches only as far as our telescopes can see.' The observed universe is a part of the whole universe and it spans thousands of Megaparsecs in space and thousands of millions of years in time. Large instruments and ingenious techniques are used to observe its structure at large scale, its content and evolution.

The investigation of the universe at large is the subject of cosmology. The approach to the large scale structure of the universe is either empirical (observational cosmology) or rational (theoretical models of the universe). The empirical method stresses observational data and aims at the real structure of the universe. The practical experience with the universe at large is, however, considerably limited. The rational method explores all possible structures of the universe – i.e., cosmological models. 'Possible' means logical, contradicting neither physical laws nor observational evidence. Mathematics greatly facilitates logical deductions and observational data are probing stones for theoretical models. There is one real universe, observed by our telescopes, and a multitude of its theoretical models. Due to the scarcity of reliable observational data about the universe at large, it is difficult to identify any particular model with the real universe.

4.9.1. Observations

The observational data concern the position, intensity of radiation (expressed in magnitudes), the spectrum and polarization of galaxies, especially of the brightest galaxies in clusters. The brightest galaxies in rich clusters can be measured at very large distances and their luminosities L are approximately equal. These two properties are used to measure distances of clusters of galaxies. For the magnitudes m_1 and m_2 of the brightest galaxies in two clusters, which are at distances r_1 and r_2 from us, the Pogson equation is

$$\frac{L}{r_1^2} : \frac{L}{r_2^2} = 10^{-0.4(m_1 - m_2)} \quad \text{or} \quad \frac{r_1^2}{r_2^2} = 10^{0.4(m_1 - m_2)},$$

which means

$$m = \text{const.} + 5 \log r. \tag{4.18}$$

The constant depends on units of distance used. It may be determined by applying Equation (4.18) to galaxies with r determined by some method used for nearby galaxies. The relation (4.18) allows us to find distances r of distant clusters by measuring the apparent magnitude m of their brightest galaxy.

Another observational characteristic of distant galaxies is the displacement $\Delta\lambda$ of spectral lines towards longer wavelengths with respect to laboratory wavelengths λ_0. The redshift of spectral lines is caused by recession of galaxies:

$$\frac{\lambda - \lambda_0}{\lambda_0} = \frac{\Delta\lambda}{\lambda_0} = z = \frac{v}{c} \quad \text{(for small } z\text{)}, \tag{4.19}$$

where v is the velocity of recession and λ is the measured wavelength in the galaxy spectrum. The redshift $\Delta\lambda$ may be sometimes twice as large as λ_0 and it is more convenient to use $(1 + z)$ for the factor by which the wavelength is increased $(1 + z = \lambda/\lambda_0)$ as follows from Equation (4.19). For high velocities v of recession the classical Doppler shift formula (4.19) cannot be used. In special theory of relativity the Doppler effect has the form

$$1 + z = \left(\frac{c + v}{c - v}\right)^{\frac{1}{2}}. \tag{4.20}$$

For $z = 2$ (which is the case with the quasar 3C9) the relativistic velocity according to Equation (4.20) would be $0.8c = 240\,000 \text{ km s}^{-1}$. But even the special theory of relativity represents an approximation. The properties of the space, described by the metrics of general theory of relativity have to be taken into account. By the presence of matter and fields the space is curved and an approximation by the flat space of special relativity is permissible only in a tiny volume, and not along the whole trajectory of the light ray from the distant galaxy.

(a) *Hubble law*

Between the two observed quantities m and z for the brightest galaxies in clusters there exists an empirical relation

$$m = \text{const.} + 5 \log cz, \tag{4.21}$$

discovered first by Hubble. As m is a measure for distance (Equation (4.18)), the empirical relation (4.21) shows the proportionality between the distance r of a cluster of galaxies and its redshift z or its velocity of recession v:

$$v = zc = H_0 r. \tag{4.22}$$

This simple relation – called Hubble law – between the distances of clusters of galaxies and their velocities of recession is valid for distances sufficiently large (for $zc \gtrsim 4000 \text{ km s}^{-1}$) so that the local effects may be neglected; but not too large ($z \lesssim 0.1$) for the relativistic effects (deceleration) to be negligible. The deviations of the very large red shifts from the simple Hubble law are of great importance, because they would decide which is the best from the cosmological models proposed. Unfortunately, in the very distant past time the luminosity L of the brightest galaxy in clusters was not necessarily the same as it is at present.

The constant H_0 – called Hubble constant – is a measure of the expansion of the universe. It is determined by observations, using pairs (z, r), where r is determined photometrically. Galaxies with velocities significantly larger than the local velocities (i.e., with $cz > 4000 \text{ km s}^{-1}$) have to be used. But there, the precise distance indicators are below the photographic plate limit. One must proceed stepwise: Cepheids in the nearest galaxies are used to calibrate brighter objects (distance indicators) such as H II regions, which in turn are used to determine the luminosity of still brighter indicators (integrated luminosity), which already reach the region of significant expansion velocities. It is not an easy task and the uncertainty of the value of H_0 thus deduced may be understood. It ranges from 55 to $100 \text{ km s}^{-1} \text{ Mpc}^{-1}$ according to different papers. The older determination gave even higher values for H_0. When later more precise determinations lead to low values of H_0, all cosmological distances (i.e., distances determined from redshift by Equation (4.22)) had to be correspondingly increased.

The reciprocal value of the Hubble constant has the dimension of time and is called Hubble time:

$$H_0^{-1} = \frac{r}{v}. \tag{4.22}$$

If the recession velocity were constant during the whole evolution of the universe, then the Hubble time would represent the age of the universe. Thus for $H_0 = 55 \text{ km s}^{-1} \text{ Mpc}^{-1}$ the Hubble time is 18×10^9 yr. But the expansion of the universe is decelerated by self-gravitation. If so, the Hubble constant was larger in the past and the Hubble time shorter than would correspond to constant velocity of expansion. The expansion of the universe started approximately 10×10^9 yr ago according to the Friedmann models and the Big Bang Theory (4.9.2(c) and Section 5.2).

4.9.2. THEORETICAL MODELS OF THE UNIVERSE

Different cosmological models have been proposed in the past, to explain the large scale structure and evolution of the universe. Here we shall mention only models based on the general theory of relativity. The theory uses a geometrical language (Section 2.4.4) in the sense that the presence of matter determines the metrical properties of the space-time. Great mathematical difficulties involved are substantially reduced by the assumption that the universe on a large scale is spatially isotropic and homogeneous (isotropic + homogeneous = uniform). The assumption of the uniformity of the universe at large scale is called the cosmological principle.

(a) *Cosmological Principle*

The principle states that the universe presents the same aspect from every place, if local irregularities are neglected. The picture of the universe for observers in any galaxy should be the same as for us. They will see other constellations, other groupings of galaxies, but the large-scale properties such as expansion of the universe, Hubble constant, deceleration parameter, mean density should be – according to the cosmological principle – the same everywhere. There is no privileged galaxy in the universe; the capital G in 'Galaxy' is only a question of orthography, not of importance.

The cosmological principle is based on observations. Hubble, Shapley and their colleagues observed that the number of galaxies to a limiting magnitude varies directly with the volume of space corresponding to the magnitude. Their additional observation that galaxies are isotropically distributed across the sky led to the conclusion that the universe is uniformly (i.e., homogeneously and isotropically) filled with matter.

(b) *Uniform Universe*

For the uniform universe the metric (Equation (2.63)) has a simple form:

$$ds^2 = c^2 dt^2 - R^2(t)\left[\frac{dr^2}{1-kr^2} + r^2(d\theta^2 + \sin^2\theta\, d\varphi^2)\right]. \tag{4.23}$$

This metric was first used by the Russian mathematician Friedmann. It was rigorously deduced by Robertson and Walker. The formula (4.23) is therefore often called the Robertson–Walker metric. The variables r, θ and φ are spherical coordinates. The cosmic time t is the same for the whole universe.

The symbol $R(t)$ denotes an unspecified function of time called expansion factor, space-dilatation factor, scale factor or space-dilatation function. It determines the curvature radius of the three-dimensional space (if $k \neq 0$), or simply the distance of two clusters of galaxies (if $k = 0$). Its meaning is analogous to the radius of curvature of a two dimensional surface. It is the same for the whole universe and only depends on the time. As a coefficient in the metric the space-dilatation function is determined by the Einstein field equations. For the uniform universe the Einstein equations reduce to differential equations for $R(t)$ (4.24 and 4.25).

The constant k in the Robertson-Walker metric (Equation (4.23)) is the index of curvature. For $k = 0$ the metric corresponds to flat euclidean space. In particular the surface of a sphere with radius r has an area $4\pi r^2$. For $k = +1$ the space is spherical, closed, and of a finite volume. Concentric spheres in this space increase their surface with increasing radius to a maximum, and beyond it the surface decreases with increasing radius to zero. The surface has an area smaller than $4\pi r^2$. The volume of the space is finite, equal to $\pi^2 R^3$, so that it increases with increasing $R(t)$. In the case of $k = -1$ the metric (4.23) characterizes hyperbolic space (called also Bolayi–Lobachewski space). The space is open and infinite. Concentric spheres in the hyperbolic space have such a property that with increasing radius increases the corresponding surface area, which is greater than $4\pi r^2$.

The geometry of the cosmic space, i.e. the value of k for the real universe, depends upon the mean density of matter and energy contained in it. In the general theory of relativity the space is curved not only by the mass of elementary particles but also by their energy and fields. The contribution of thermal energy, kinetic energy of stars and galaxies, of magnetic fields, of photons etc. is at present very low so that the rest mass of particles represents practically the whole mean density ϱ_0 of the present universe. Quite an opposite situation was in the radiation era of the universe (Section 5.2).

(c) The Actual Universe

From our Galaxy the universe appears uniform on a large scale. The assumption of uniformity (i.e., cosmological principle) is thus justified. The properties of the uniform universe are described by a simple Robertson-Walker metric (4.23). The metric characterizes spaces of constant curvature everywhere in the universe at a given moment. The dilatation function $R(t)$ and the curvature constant k are not specified in Equation (4.23) because the metric is valid for any uniform universe.

When the Einstein field equations are formulated for the general metric (4.23), one finds that the original ten equations reduce to

$$\frac{\dot{R}^2}{R^2} + \frac{kc^2}{R^2} = \frac{8\pi G\varrho}{3} \tag{4.24}$$

and

$$\frac{2\ddot{R}}{R} + \frac{\dot{R}^2}{R^2} + \frac{kc^2}{R^2} = -\frac{8\pi Gp}{c^2}, \tag{4.25}$$

where ϱ and p are the mean density of the actual universe and its mean pressure (or energy density, negligible at present). $\dot{R}$ and $\ddot{R}$ are the first and second time derivatives of the dilatation function.

Observers prefer to use the first and second time derivatives of the dilatation function in the form of

$$\text{Hubble constant } H = \frac{\dot{R}}{R} \tag{4.26}$$

and

$$\text{deceleration parameter } q = \frac{-R\ddot{R}}{\dot{R}^2}. \tag{4.27}$$

In order to determine the geometry of the actual universe, we substitute Equations (4.26) and (4.27) in the field equations (4.24) and (4.25) (with $p = 0$ in the last equation, because pressure is negligible at present) to obtain

$$k = \frac{R^2}{c^2}\left(\frac{8\pi G\varrho}{3} - H^2\right) \tag{4.28}$$

and

$$q = \frac{4\pi G\varrho}{3H^2}. \tag{4.29}$$

Substituting the deceleration factor (Equation (4.29)) into the curvature constant k (Equation (4.28)) one gets

$$-k = \frac{R^2}{c^2}H^2(1 - 2q). \tag{4.30}$$

As R^2/c^2 is positive, the formulae (4.28) and (4.30) imply:

$$\left.\begin{array}{l} \text{for flat space,} \qquad k = 0 \Leftrightarrow \varrho = \dfrac{3H^2}{8\pi G} \Leftrightarrow q = \tfrac{1}{2} \\[3mm] \text{closed (spherical) space,} \; k = +1 \Leftrightarrow \varrho > \dfrac{3H^2}{8\pi G} \Leftrightarrow q > \tfrac{1}{2} \\[3mm] \text{open (hyperbolic) space,} \, k = -1 \Leftrightarrow \varrho < \dfrac{3H^2}{8\pi G} \Leftrightarrow q < \tfrac{1}{2}. \end{array}\right\} \tag{4.31}$$

The implications of Equation (4.31) mean that from two observable quantities (the Hubble constant and deceleration parameter) the geometry of the universe may be determined and it may be decided whether the universe is flat ($k = 0$), or open and expanding forever ($k = -1$) or closed with decelerated expansion which will stop and reverse into contraction ($k = +1$) (Figure 4.31).

It is remarkable that only two quantities have to be determined by observations with large telescopes, in order to find out the mathematical (Friedmann) model corresponding best to the actual universe. The Hubble constant defined by the relation (4.22) determines the present expansion rate and the deceleration parameter corresponds to the change of the expansion rate with time. The two numbers H and q define the Riemann curvature c^2/R^2 (4.30) of the cosmic space at a constant time. The two values tell us whether our universe is ever-expanding or whether it performs oscillations involving expansion and then contraction (Equation (4.31)). The mean density of the present universe may also be predicted from the two values (Equation (4.29)). The time t_0 from now to the start of expansion ($t = 0$) is a known function of H and q (Friedmann time, age of the universe). For the closed (spherical) universe the duration of the whole cycle expansion-contraction may be deduced from the observed H and q values:

$$T = 2\pi q H^{-1}(2q - 1)^{-3/2}. \tag{4.32}$$

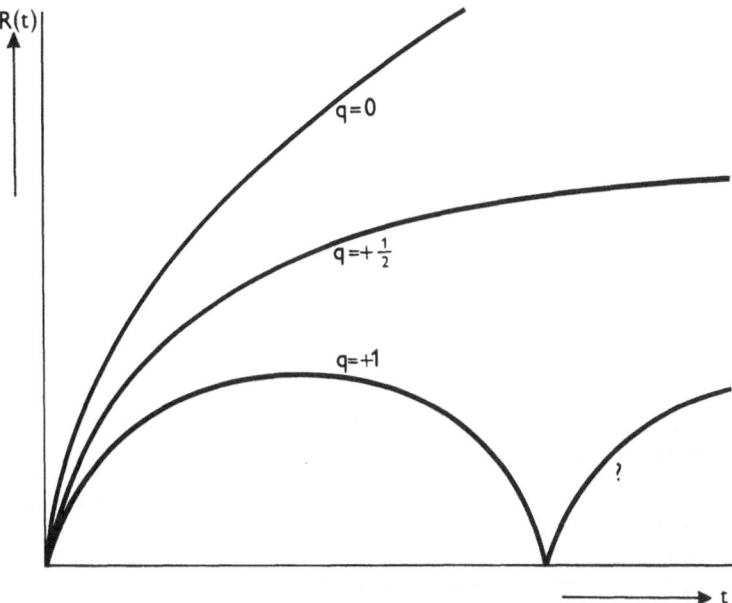

Fig. 4.31. The geometry (i.e., expansion function $R(t)$) is given for three different values of the deceleration parameter q. For the upper curve ($q = 0$, i.e., $k = -1$) the universe will expand forever. For the lower curve ($q = +1$, i.e., $k = +1$) the universe is closed. For $q = \frac{1}{2}$ ($k = 0$) the universe is infinite and euclidean.

Though the exact values of H and q are not known, one can adopt $H = 75 \ \mathrm{km \ s^{-1} \ Mpc^{-1}}$ and $q = +1$; then the total cycle time is $T = 8 \times 10^{10}$ yr (Figure 4.32).

The critical density to close the universe, viz.

$$\varrho_c = \frac{3H^2}{8\pi G} \tag{4.33}$$

follows from Equation (4.31). The critical density determined from the Hubble constant is more than one order of magnitude higher than the density of luminous matter in galaxies. (However, the uncertainty of the H value must be stressed here.) The missing mass (and energy) is supposed to exist mainly in the intergalactic space in a non-luminous form: cold intergalactic H, dark discrete objects until now undetected (evolved galaxies, black holes, black dwarfs), neutrinos, relativistic particles, electromagnetic fields, gravitational radiation etc. However, there is no compelling observational evidence for the presence of significant amounts of mass and energy besides galaxies. The problem of the missing matter remains open.

It is also possible that our universe is open, that the actual mean density is smaller than the critical density of the model, i.e. $\varrho < \varrho_c$. Then there would be no problem with the missing matter. But another problem would appear, i.e., why the deceleration parameter is not smaller than $\frac{1}{2}$ as should be the case for the open (hyperbolic) universe (see Equation 4.31) last row). The present observations lead to $q = +1$ (Figure 4.33) and indicate that the actual universe is probably closed.

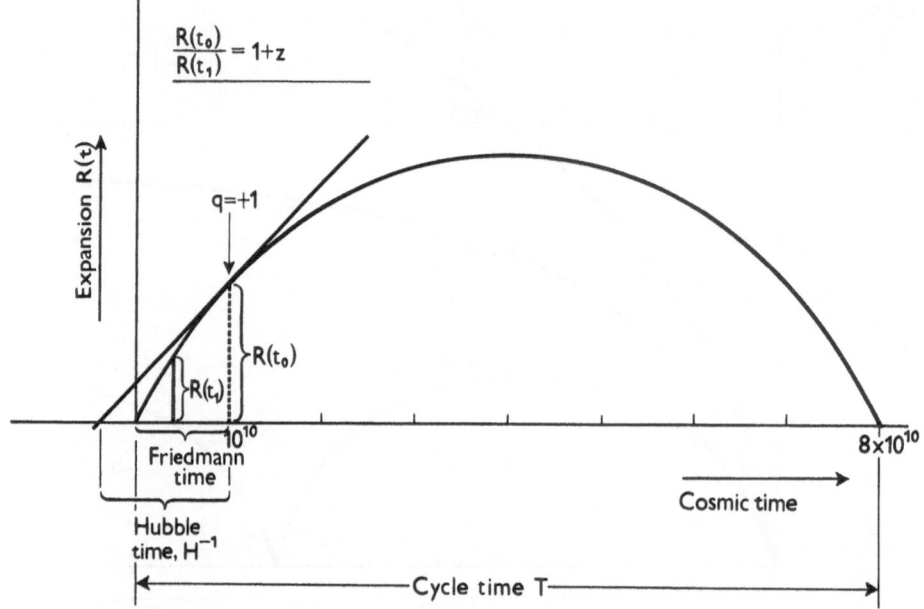

Fig. 4.32. Cycle of the closed universe for the deceleration parameter $q = +1$.

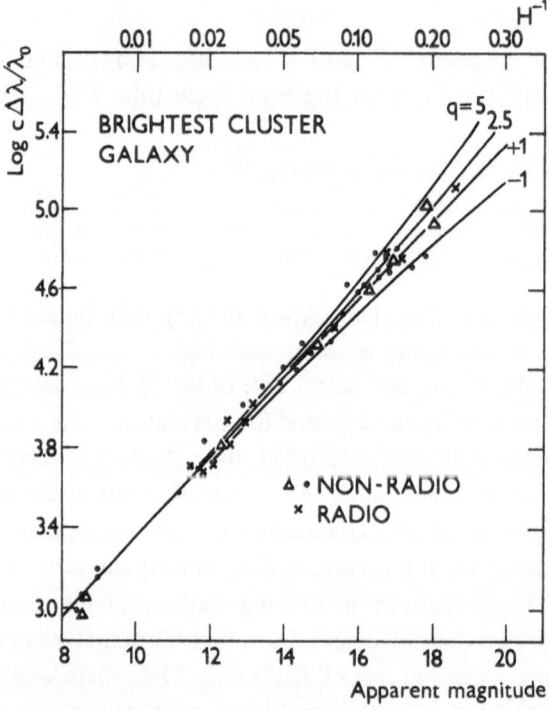

Fig. 4.33. Relation between redshift and apparent magnitudes of galaxies. Brightest galaxies in 42 clusters are plotted. Lines computed for different values of the deceleration parameter are superposed (A. Sandage). The observations seem to fit the curve for $q = +1$.

Summary

Matter in the universe is agglomerated in objects of different mass, size, structure and lifetime – ranging from elementary particles to clusters of galaxies. The objects (or better the classes of objects) may be arranged according to the number of constituting nucleons (i.e., according to their mass). Whatever the complexity of an object may be, its structure is based on elementary particles and their interactions.

Protons, electrons and neutrons are the most important particles for the structure. Though they are called elementary, it is not clear whether they themselves do not consist of more fundamental entities (quarks, partons).

Atomic nucleus is a system of protons and neutrons held together by strong interactions. Atomic nuclei are produced in stars (nucleosynthesis) and their liberated binding energy is the main source of stellar radiation. Though the binding energy of nucleons is high (≈ 8 MeV per nucleon) the nucleus may be destroyed by very high temperature (as in supernovae), by cosmic radiation (e.g., in interstellar space, in terrestrial atmosphere etc.) or by neutronization under very high pressures (e.g., during the collapse of a neutron star).

The space of the nucleus is closed in respect to strong interactions, but it is open for electromagnetic and gravitational forces. The positive charge of the nucleus binds negative electrons to form a neutral atom. Only a very small fraction of nuclei in the universe can afford an electronic envelope: the major part of matter in the observable universe is concentrated in hot stellar interiors and there the most abundant elements (H and He) are completely ionized. The binding energy of electrons in the atoms is 10^5 to 10^6 lower than the binding energy of nucleons in a nucleus. Atoms are therefore considerably less stable than nuclei and they can exist only in relatively narrow intervals of temperatures and densities.

The electron envelope does not cancel electric forces of the nucleus completely even if the number of electrons is equal to the number of protons. The residual electric forces give rise to chemical bonds which unite atoms to molecules. Residual electric forces of molecules (Van der Waals forces) organize higher structures: molecules to crystals, crystals to rocks, meteoroids, small satellites and small asteroids. With the increasing size of structures the role of the weakest interaction (gravitation) increases until it becomes dominant in planets. Their spherical figure corresponds to minimum potential energy while small bodies held together mainly by electromagnetic forces have irregular shapes (Phobos, Deimos, small asteroids).

The largest single bodies in the universe are stars. A star is a system of 10^{56} to 10^{58} nucleons and electrons bound together by their mutual gravitational attraction (self-gravitation). Stars represent fundamental structural units in the hierarchy of the universe because: (1) most of the observable matter is concentrated in them; (2) they process the cosmic material by thermonuclear reactions and are thus responsible for chemical aging of the universe; (3) they themselves are fundamental units in the structure of higher systems (planetary systems, multiple stars, stellar clusters, galaxies).

Gravitation is the main force to keep the star together but the other three interactions are also important for the structure of stars. Yet it is the only force governing structure and evolution of stellar systems. The same is true for the structure and evolution of groups and clusters of galaxies.

The universe at large scale is expanding so that the distances between its structural units (i.e. clusters of galaxies) are increasing with time. The expansion is slowing down by mutual gravitational attraction between clusters. The observed deceleration q and the rate of expansion H indicate that our universe started its expansion (Big Bang) about 10×10^9 yr ago (Friedmann time or age of the universe). The estimated value $q = +1$ means that we live in a closed finite universe the whole cycle of which (i.e., expansion from Big Bang to a standstill and recontraction to singularity) should last 8×10^{10} yr. With its 10^{10} yr of age our universe appears still relatively young.

CHAPTER 5

EVOLUTION

The structure of the universe discussed in the preceding chapter is a consequence of past events. There are various pieces of evidence that the universe in the past was different from its present structure: chemical composition of young stars differ from that of old stars, the relic microwave radiation with present temperature 3 K whose temperature continues to decrease with time, counts of radio sources, expansion of the universe etc.

The whole hierarchy of the observed structures originated and evolved in the span of 10^{10} yr from the Big Bang up to the present time. It was a history of all the existing matter – the true *theatrum mundi* – in frame of which any other history (e.g. history of mankind, of the whole terrestrial biosphere or of any other galactic community) appears as a tiny episode.

5.1. Evolutionary Processes

The astronomy is 10^7 times younger than the age of its subject – the universe. From the times of Hipparchos two thousand years ago, the sky remained nearly unchanged. There are practically no written documents concerning the changes of stars (except about a few supernovae). The reason is evident: the evolution of stars and other cosmic bodies proceeds much more slowly than the evolution of human civilization. But for all that, the astronomers and physicists accumulated – particularly during the last decade – a large amount of data on evolutionary processes.

5.1.1. EVOLUTIONARY PROCESSES (SMALL AND LARGE SCALE)

There is a variety of processes that shaped the amorphous hot matter of the fireball and transformed it gradually into the rich structural hierarchy of the present universe. Let us mention some of them:

Formation (materialization) of elementary particles and their decay or complete annihilation into γ photons; emission of photons when the kinetic energy of charged particles is transformed into electromagnetic radiation; emission of neutrinos by β^+-decay of protons in nuclei which are synthetized by thermonuclear reactions in stars.

Synthesis of nuclei in stars by different processes – pp-chain, CN-cycle, capture of charged particles, slow neutron capture and rapid neutron capture; equilibrium processes by which the most stable elements (Fe group) are synthetized; spallation reactions in interstellar space giving rise to Li, Be, B.

Processes in the electron envelope of atoms; excitation, deexcitation; emission and absorption of photons; ionization, recombination and electron attachment; a considerable part of interaction between matter and the photonic component of the universe proceeds via the electron envelope of atoms or ions.

Formation of molecules and grains wherever conditions are favorable; simpler organic molecules are synthetized in interstellar clouds in large quantities; complex prebiotic molecules have been synthetized in our planetary system and they are extracted from carbonaceous chondrites; organic compounds are probably formed in the Jupiter atmosphere; at the surface of our planet the molecular synthesis advanced to such a degree that life could originate.

Formation of planets and all the other members of the planetary system down to the tiniest micrometeoroids; various processes that shaped surfaces of planets and their atmospheres; origin, sublimation and dissolution of cometary nuclei; encounters and collisions of meteoroids with asteroids (fragmentation), impact of meteoroids, asteroids or comets on surface of planets and satellites (meteors, bolids, meteorites, meteoric dust, meteoritic dust and impact craters); perturbation of orbits of planetary system members by gravitation of giant planets; sublimation of meteroids near the Sun; sweeping out of the smallest micrometeoroids from the solar system by pressure of solar radiation; breaking-up of asteroids etc.

Gravitational contraction of protostars, production of energy in the stars and its transport to the surface and into the interstellar space; thermonuclear reactions and changes of stellar structure; interaction of stars with interstellar medium; accretion and ejection of matter by stars; stellar winds; changes of stellar rotation; variability of stars; stellar activity (spots, flares, prominences, coronae); over-spilling of plasma from one component of close binaries on the other as in some recently discovered X-ray sources; gravitational collapse of stars in supernova event; transition of stellar matter from the state of plasma to degenerate gas (white dwarfs) or neutron gas (neutron stars); cooling of stars without energy sources (white dwarfs and very light red dwarfs, black dwarfs); slowing down of rapidly spinning pulsars; feeding of energetic electrons into supernova remnants; acceleration of ions to relativistic energies (cosmic rays) etc.

Formation of stellar associations and stellar clusters; their decay and disruption by galactic gravitation; deformation of main sequence of stellar clusters as an age indicator; ejection of stars from clusters etc.

Formation of galaxies; activity of their nuclei; formation of spiral arms; interaction of close galaxies; expansion of the universe at large scale and deceleration of the expansion rate.

The complete list of all the evolutionary processes that shape the universe could be lengthened and the reader could improve the incompleteness by adding other evolutionary processes to the list. More impressive than the length of list of the evolutionary processes in the universe is, however, the fact that all the processes are the result of one or more fundamental interaction (i.e., gravitational, weak, electromagnetic and strong) between the constituent elementary particles.

5.1.2. COSMOCHRONOLOGY

Cosmochronology has the purpose to arrange the important evolutionary processes into a time sequence. It tries to provide the evolution of the universe with a time scale. Different phenomena may be used for age determination:

(a) *Expansion*

From the expansion rate (Hubble constant) and its deceleration (deceleration constant) the start of the expansion is computed. It is the age of the universe, called Friedmann time.

The expansion of supernova remnants (e.g. of Crab nebula) may be used to find out when the parent supernova exploded. One divides simply the radius of the remnants by velocity of expansion. In the case of the Crab nebula the expansion age is confirmed by historical documents.

Still another example is the dating of explosive events in galactic nuclei using velocities and distances of their ejecta. Thus the nucleus of the galaxy M 82 exploded about 1.5×10^6 yr ago.

(b) *Thermonuclear Reactions*

In the main sequence stars H is transformed into He. Having exhausted completely the H in its central part and formed a helium core, the star turns off from the main sequence because its He core representing about 12% of the star's mass, contracts and its envelope expands. The star moves away from the main sequence into the red giant region. The massive and luminous stars in the upper left part of the main sequence convert H very fast due to their high central temperature. Less massive stars (situated on the lower part of the main sequence) stay immobile much longer on the main sequence. Theory of internal stellar structure predicts the duration of the stay in each point of the main sequence. The stars in stellar clusters have the same chemical composition, the same age and they differ only in their initial mass. From the Hertzsprung-Russell diagram of a stellar cluster the turn-off point may be found and this point permits an easy determination of the age of the cluster. The ages determined in this way range from a few millions of years (for young open clusters) to about 10^{10} yr (for globular clusters).

(c) *Natural Radioactive Isotopes*

Half-lifetimes of radioactive isotopes are independent of what happened to them since they were formed. They offer a useful method of dating events in the universe. Well known examples are the determination of ages of terrestrial rocks, meteorites and of lunar samples. After a rock has solidified, the radioactive elements (such as ^{232}Th, ^{235}U and ^{238}U) remain associated with the daughter isotopes of Pb (^{208}Pb, ^{207}Pb and ^{206}Pb) to which they decayed. In solid state they could not be separated from their parent isotopes. Their relative abundances depend upon the decay rates (half lifetime) and upon the time since the rock solidified (solidification age). These decay rates are well known. The meteorite age estimates lead to 4.6×10^9 yr. The oldest rocks on the Earth are 3.7×10^9 yr old. The age of the Earth determined from the isotopic abundances of the primordial Pb is 4.6×10^9 yr. The ages of certain lunar sample 'times' are practically the same and we may conclude that the solid state in our planetary system is about 4.6×10^9 yr old.

Radioactive isotopes and their decay products (in the planetary system) may be used to obtain information about history of nucleosynthesis in our Galaxy, before the formation of the planetary system. The radioactive isotopes were at that time produced in supernova explosions by r-process of neutron capture. Their relative original abundances are computed from their production rates. Comparison with the present empirical abundances indicate that the radioactive isotopes were synthetized between 0.7×10^{10} and 1.5×10^{10} yr ago.

It is remarkable that the age of the universe determined by the three quite different dating methods gives approximately the same value, about 10^{10} yr (or slightly more).

5.2. Big Bang

Our universe is expanding at the present time. In theoretical models the rate of expansion is described by the scale function (expansion factor) $R(t)$ defining the same metric in the whole universe as a function of universal time t. Empirically the expansion is confirmed by redshift of distant galaxies (Hubble law), and by the darkness of the night sky (Olbers paradox).

5.2.1. HOT SUPERDENSE UNIVERSE

The start of the expansion – about 10^{10} yr ago – is usually called Big Bang. The function $R(t)$, which determines distances between two clusters of galaxies, is an increasing function of the age of the universe t. Therefore, looking deeper into the past (smaller t), one finds $R(t)$ smaller (Figure 4.32) and the mean mass density of the universe ϱ_m to become larger in accordance with

$$\varrho_m \propto R(t)^{-3} \tag{5.1}$$

Today, the observed mean density is about 3×10^{-31} g cm^{-3} or 3×10^{-10} erg cm^{-3}. Near the start (t close to zero) the universe was superdense.

The expansion of the universe also lowers the density of its photon component. Not only the number of relic photons per cm^3 decreases with time $R(t)^{-3}$ but also the energy of each photon decreases as

$$h\nu \propto R(t)^{-1}. \tag{5.2}$$

It means that the density of relic radiation ϱ_r changes by expansion as

$$\varrho_r \propto R(t)^{-4} \tag{5.3}$$

i.e., faster than the mass density of particles. The relation (5.3) simply follows from the adiabatic expansion of a photon gas in an expanding perfectly reflecting enclosure. Its volume V and pressure p are related as

$$pV^\gamma = \text{constant} \quad (\gamma = \tfrac{4}{3}). \tag{5.4}$$

Volume V expands with time as $R(t)^3$, so that $p \propto R(t)^{-4}$. The equation of state for photon gas is

$$p = \frac{\varrho_r c^2}{3}, \tag{5.5}$$

from which Equation (5.3) follows. In the present universe ϱ_r is about 10^{-33} g cm^{-3} (10^{-12} erg cm^{-3}) mainly in the form of relic radiation.

Going backwards to smaller t, the radiation density ϱ_r of the universe increases faster than the mass density ϱ_m of matter, until at certain times in the past both were equal. Before that time, in the more remote past, the radiation was more important than material particles (the so-called radiation era of the history of universe). Due to Equations (5.2) and (5.3) the universe was hotter when it was young (smaller t). Near the start of expansion when t was close to zero, all distances were close to zero (Friedmann singularity) and the universe was extremely hot and dense. Its state would correspond to a point far outside the upper right corner of Figure 3.1. The primordial substance is sometimes called ylem, cosmic ylem, primordial fireball, primeval fireball, hot universe, primeval atom, polyneutron cosmic liquid etc., by different authors.

In the initial stages of the expanding universe

$$\varrho_r \gg \varrho_m \quad \text{so that} \quad \varrho = \varrho_r + \varrho_m \approx \varrho_r = \frac{aT^4}{c^2} \tag{5.6}$$

where aT^4 is the radiation density (Stefan law) and the constant a is

$$a = \frac{8\pi^2 k^4}{15 c^3 \hbar^3} = 7.6 \times 10^{-15} \text{ erg cm}^{-3} \text{ deg}^{-4}. \tag{5.7}$$

For the density in the early phases of universe we have from Equations (5.6) and (5.7)

$$\varrho = \frac{aT^4}{c^2} \approx 10^{-35} \, T^4 \approx \frac{kT}{c^2}\left(\frac{kT}{\hbar c}\right)^3, \tag{5.8}$$

where $kT \approx h\nu$ corresponds to a mean energy of a photon, kT/c^2 is its equivalent mass, so that $(kT/\hbar c)^3$ would then be number of photons in cm^3.

The function $R(t)$ may be deduced from the present expansion rate H and its deceleration q. If one uses $R \propto \sqrt{t}$, then Equation (5.3) becomes

$$\varrho \approx 10^6 \, t^{-2} \quad (\varrho \text{ in g, } t \text{ in s}). \tag{5.9}$$

The relations (5.8) and (5.9) determine the way in which the temperature of the

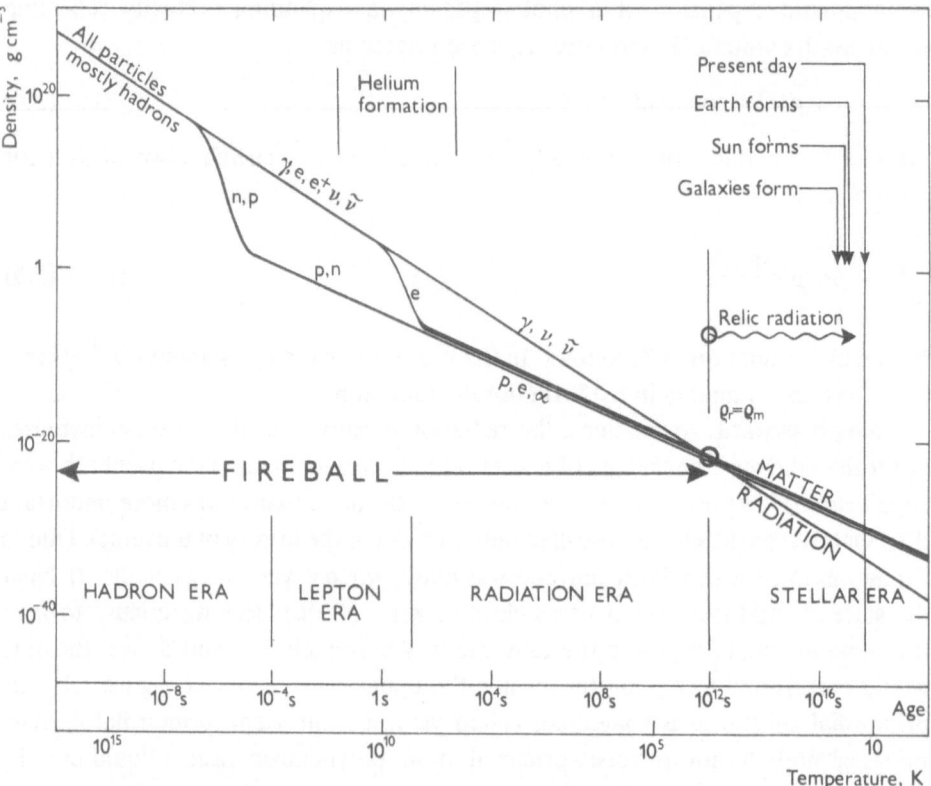

Fig. 5.1.   Evolution of particle and photon components of the universe. Abscissae have double scale: age of the universe and the corresponding temperature. On the ordinate axis is the density for both components. The moment when densities of both components are equal ($\varrho_r = \varrho_m$) is the end of the radiation era and the beginning of the stellar era. The period of He synthesis is marked.

expanding universe changed with time, namely

$$T \approx \frac{10^{10}}{\sqrt{t}} \, \text{K}. \tag{5.10}$$

5.2.2. Chronology of the Universe

The relations (5.9) and (5.10) define the density and temperature of the expanding universe as functions of its age (Figure 5.1). They and the relation $h\nu \approx kT$ make it possible for us to study the earliest phases of the expanding universe. With temperature decreasing in accordance with Equation (5.10) the photon energy decreased so that the process of materialization was gradually limited to particles with lower and lower rest mas ($h\nu \sim m_0 c^2$). With respect to this gradual limitation of materialization process, the chronology of the universe is divided into four eras. It is represented in Figure 5.1.

(a) *Hadron Era*

In the beginning of the expansion the temperature was very high, so that all particles existed (Section 3.6) in approximately equal amount. As there are many species of hadrons (Figure 1.3) (leptons exist only in three species), the main constituents of the exploding universe in the first moments were hadrons and hence the name of era. The γ-photons had sufficient energy to materialize (marked by the lower arrow):

$$\text{baryon} + \text{antibaryon} \rightleftarrows \text{gamma photons} \tag{5.11}$$

and due to high densities the pair immediately annihilated (the upper arrow in Equation (5.11)). The number of baryons was nearly equal to the number of antibaryons, it was only by a fraction 10^{-9} larger! That means that only one among 10^9 baryons had no partner with which to annihilate itself. This very slight asymmetry in favor of baryons was important for the future of the universe. After the first millionths of a second ($t \sim 10^{-6}$ s) the temperature of the expanding universe dropped to $\sim 10^{13}$ K (according to Equation (5.10)) with a mean energy of particles and photons $\sim 10^3$ MeV, which is the rest energy of baryons. Later on, the formation of baryon-antibaryon pairs stopped because the photons were not energetic enough. The annihilation (Equation (5.11)) soon depleted all antibaryons from the hot expanding universe.

Only those baryons remained which had nothing to annihilate themselves with (10^{-9} of the original number). They cascaded to the lowest state, i.e. to nucleons. When at $t = 10^{-4}$ s the temperature dropped to 10^{12} K, the energy of particles and photons became too low (100 MeV) for the production of π-mesons ($T < m_\pi c^2/k$), the π-mesons died out and new ones could not be formed. The π-mesons are the lowest-mass hadrons (i.e., strongly interacting particles), so that the hadron era ends at the time $t = 10^{-4}$ s. Never after would the strong interactions assert themselves in such a high degree as in this most fascinating period of the universe, which lasted from 10^{-44} to 10^{-4} s, cooled from 10^{32} K down to 10^{12} K, and expanded from 10^{94} g cm^{-3} to 10^{14} g cm^{-3}. For $t \approx 10^{-44}$ s, $\varrho \approx 10^{94}$ g cm^{-3} and $T \approx 10^{33}$ K due to quantum fluctuations of the metric, the classical cosmology becomes meaningless.

(b) *Lepton Era*

When the energy of photons and particles dropped from 100 MeV to 1 MeV, leptons were very abundant. The temperature was sufficient for copious formation of electrons, positrons and neutrinos. Baryons, which survived from the hadron era, were rare ($\sim 10^{-9}$) in comparison with the leptons and photons.

The lepton era begins with the decay of the last hadrons – pions – into muons (Equations (2.48) and (2.49)). It ends after a few seconds at 10^{10} K with electron–positron pair annihilation (Equations (1.15) and (2.39)). During this era the neutrinos had less and less to do with the rest of matter until they decoupled completely and began an independent existence. The neutrinos, called relic neutrinos, should fill the universe – similarly to relic microwave radiation.

At the beginning of the lepton era the number of neutrons was equal to the number of protons. Later on the neutrons decayed (Equation (2.19)) with half-life of 12 min, with the effect that the number of protons increased to the detriment of neutrons. But, apparently before they could all decay, many neutrons combined with protons to form deuterium $(n+p \rightarrow {}^2H+\gamma)$ which was followed by formation of He nuclei (Figure 5.2). The photon energy $\sim 10^5$ eV was below the binding energy of the nuclei, so that they could not be destroyed. How much of the deuterium and He in the present universe dates from the lepton era and beginning of the radiation era is still an open question.

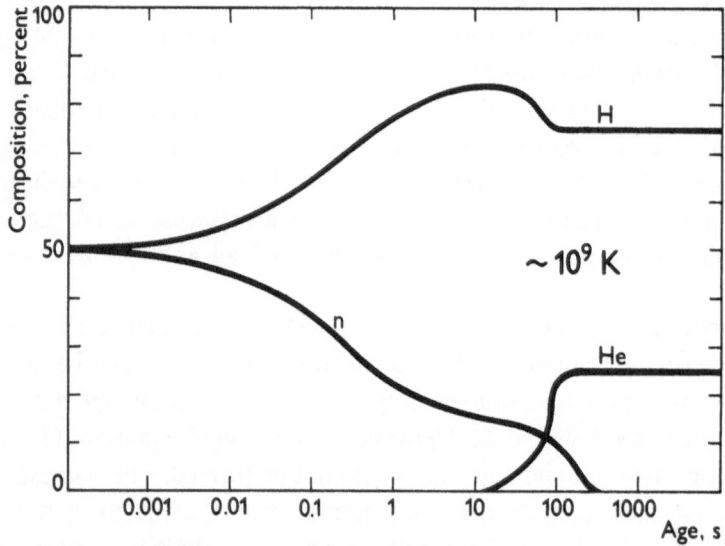

Fig. 5.2.   Neutrons produced He with protons before they could decay. The major part of the present He abundance is supposed to be of the cosmological origin.

If there had been no He synthetized in prestellar era, the stars would produce only He/H $\sim$ 3% to 5% abundance (by mass). However the abundances of He in young stars and nebulae are considerably higher (He/H $\sim$ 25 to 30% by mass); and this means that the bulk of the He should have been produced before formation of galaxies (see helium formation in Figure 5.1). Or the He was synthetized in supermassive stars during the early period of our Galaxy when the latter was much more luminous than it is today. The nucleosynthesis of He is still a problem lively discussed by astronomers.

(c) Radiation Era

The radiation era began when the temperature of the universe dropped below 10^{10} K (i.e., below 1 MeV). All electron-positron pairs were annihilated and all materialization processes had ceased. Remaining electrons, protons (and some newly formed

α-particles?) floated among the photons that were nine orders of magnitude more numerous than the particles.

The photon component of the universe prevailed over the matter (i.e., over particles that possess rest mass): $\varrho_r > \varrho_m$. Therefore the period is called the radiation era. But ϱ_r decreases faster (Equation (5.3)) than ϱ_m did (Equation (5.1)) until both become equal at the end of the radiation era. Matter emerges as the dominant component of the universe, though the ratio of photons to nucleons remained (and remains) unchanged and very high (10^9).

At the temperature of about 10^4 K, electrons recombined with protons to form neutral H. That is the end of the so-called prestellar period which comprises all the three eras discussed above. The period also called the fireball stage of the universe, ended when the universe was about 3×10^5 yr old, i.e., $1/30\,000$ of the present age. The scale factor $R(t)$ at the end of the radiation era was approximately 10^{-3} of the present value $R(t_0)$.

(d) *Stellar Era*

After recombination of H atoms the evolution proceeded independently for matter and for radiation. In particular

$$T_r \propto R(t)^{-1} \tag{5.12}$$

as follows from $\rho_r \propto R(t)^{-4} \propto T^4$ (Equations (5.3) and (5.6)). For the expanding matter with state equation $pV = RT$, the adiabatic expansion $pV^\gamma = $ const. ($\gamma = 5/3$) gives $T_m V^{2/3} \propto T_m (R(t)^3)^{2/3} = T_m R(t)^2 = $ constant, or

$$T_m \propto R(t)^{-2}. \tag{5.13}$$

As may be seen, the radiation temperatue T_r (after decoupling) decreases more slowly than the temperature of matter T_m. For densities of matter ϱ_m and radiation ϱ_r the reverse is true (Equations (5.1) and (5.3)).

The stellar era is characterized by supremacy of matter i.e., by the formation of clusters of galaxies, radio galaxies, quasars, normal galaxies, star clusters and stars, planetary systems and of all the chemical elements and molecules; the evolution of matter led to the appearance of life – certainly on one planet and probably on many others.

While the nucleons and electrons were organized subsequently by the four interactions into a great variety of systems we observe in the universe, the relic photons evolved in a much simpler manner: their density and energy simply decreased with time. At present they are recorded by radio telescopes as isotropic thermal radiation temperature $T_r = 3$ K. The density of the relic microwave photons is about 10^3 cm^{-3}, while the mean density of the observed material universe is about 10^{-6} nucleons per cm^3. Their ratio is therefore 10^9.

5.2.3. ANTIMATTER IN THE UNIVERSE

In Section 5.2.2 (a) it has been assumed that there was a slight asymmetry in favor of baryons. Such a slight asymmetry viz. $(1 + 10^{-9})/1$ would explain why the present universe seems to consist of ordinary matter ('koinomatter') only. But the asymmetry is difficult to understand and there is no decisive argument either in favor of or against the existence of antimatter in the universe. Is it possible to prove that quasars and very active galactic nuclei do not consist of ordinary matter surrounded by antimatter or vice versa? The photons emitted by antimatter and those from ordinary matter are the same.

An alternative cosmological hypothesis assumes that there was no such asymmetry as discussed above, i.e., the number of baryons equalled the number of antibaryons – so that the total baryonic number for the whole universe was and still is zero. According to this hypothesis, the nucleons and antinucleons were separated during the radiation era. Then the ordinary matter (koinomatter) and the antimatter were reorganized into increasingly larger regions of space. Annihilation along the contact surface produces pions which decay (process (1.25)) among others into γ-protons. The high energy decay products carry a large momentum and exert a strong pressure upon matter on one side and upon antimatter on the other one (Leidenfrost effect). The pressure caused by the annihilation might have been the driving force for the Big Bang explosion (Figure 5.3). The hypothesis thus escapes difficulties with the

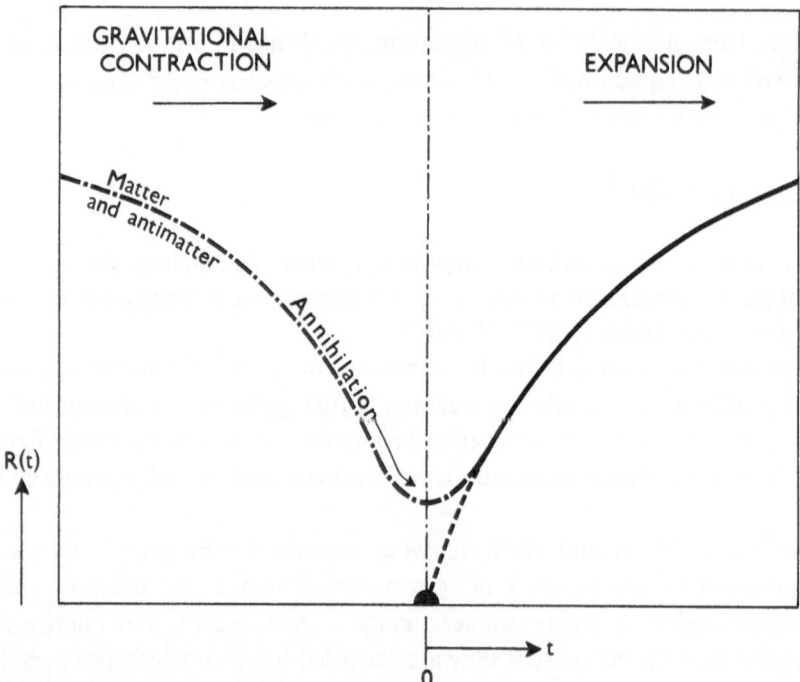

Fig. 5.3. The start of the expansion of the universe is explained (by H. Alfvén) as consequence of annihilation of matter with antimatter. The annihilation produced pressure which then reversed the gravitational contraction into expansion.

Friedmann singularity at $t \rightarrow 0$ with infinite densities and unknown properties of matter.

The Big Bang theory of the universe history, though generally accepted at present, is not yet established beyond reasonable doubts. Another well-known theory is the steady state cosmology. In contrast to the evolutionary expanding universe it assumes that the universe does not change with time and new matter is steadily created to compensate for the expansion.

5.3. Galaxies – Origin and Evolution

The cosmological theories assume that on a large scale the universe is homogeneous and isotropic (i.e., uniform). Or, in other words, there is no privileged place from which the universe would appear different from what we observe. The universe at large is homogeneous and isotropic (i.e., uniform). It is assumed that in the evolutionary universe (Friedmann model) the scale factor $R(t)$ and the curvature parameter k (or observationally the Hubble constant H and the deceleration parameter q) are the same through the whole universe at a given time.

The recent measurements of extragalactic radio sources suggest that, on a large scale ($\geqslant 10^8$ pc), the overall distribution of matter is smooth and that it is expanding in a homogeneous and isotropic way. If there is anisotropy of the radio sources, it is less than 1%. The fossil microwave radiation indicates a strict isotropy to about 0.1%. This radiation is probably a relic from the early universe and as such it provides information on isotropy over the entire path from here backwards toward the beginning itself.

5.3.1. INHOMOGENEITIES AND GALAXY FORMATION

On a (cosmologically) small scale the universe is of course not homogeneous and isotropic. Matter is clumpy on a scale of 30 to 60 Mpc downwards: it is concentrated in stars, galaxies, clusters of galaxies and the largest inhomogeneities (superclusters of galaxies) have the size of the order of 30 to 60 Mpc. The cosmology and cosmogony must explain how these density irregularities were formed.

It is likely that the unevenness on the scale of 30 to 60 Mpc originated from the large-scale inhomogeneities that were already present in the rapidly expanding fireball. Concentrations of mass $\sim 10^{15} M_\odot$ were probably separated during the prestellar stage either by a smaller expansion rate, larger densities, or both. They survived the further evolution until present times, and we observe them as clusters and superclusters of galaxies – the largest structural units of the universe. The superclusters and clusters of galaxies thus appear to be a heritage from the earliest times of Big Bang; they reflect inhomogeneities in the fireball.

Most of the galaxies were formed later within the primeval superclusters. On the scale of galaxies (mass $\sim 10^{11} M_\odot$ or less) the matter was apparently in a highly turbulent state. The primeval turbulence must have played an important role in the formation of protogalaxies; until the present time the distribution and appearance of galaxies seems like the fossil remnants of huge turbulent eddies. Many galaxies rotate fast and for some of them, located in our neighborhood, angular momentum and mass can be estimated from the measured rotation. It comes out that at the end of the radiation era when matter and radiation decoupled, the density was so high and the volume containing the mass of a galaxy so small that it could not contain the angular momentum measured at present. Some astronomers, conclude therefore, that most protogalaxies had to separate from the pregalactic substrate considerably later – when the mass of the protogalaxies expanded so much that it could contain the angular momentum. At the epoch when the protogalaxies were separated within the protoclusters as structural units, the scale of the universe was about 30 times larger than at the end of the radiation era.

5.3.2. EVOLUTION OF GALAXIES

(a) *Age of Galaxies*

There is some evidence that galaxies of the Local Group – and probably most galaxies beyond it – were formed at roughly the same time. The population of globular-cluster-like stars observed in the galaxies support the idea. Even galaxies that seem to be young, such as both Magellanic Clouds, are as old as the oldest stars in our Galaxy. The globular clusters represent the first generation of objects formed in galaxies. Their age is determined from the turn-off point where their main sequence terminates. It is about 10×10^9 yr (with rms error of 3×10^9 yr).

Although observation and theory agree that most normal galaxies were formed already about 10×10^9 yr ago, the possibility is not excluded that some galaxies were formed at a much later time, for example by ejection from massive galactic nuclei or by condensation of denser parts of intergalactic matter.

The process of star formation in galaxies has continued until the present time – as may be seen from the presence of very young stars in spiral arms. Besides formation of new stars from interstellar gas-dust clouds, the evolution of galaxies is also characterized by chemical changes of their material produced by thermonuclear reactions in stellar interiors. The structural and dynamical evolution of galaxies concerns the spatial distribution of stars, interstellar matter and the velocity of rotation around the galactic axis. The formation and evolution of stars and the chemical evolution of galaxies are closely interconnected and will be treated in Section 5.4.

(b) *Halo, Contraction and Disk Phases of the Galaxy*

The self-gravitation of the H protogalaxy (with some He admixed) started its contraction and the halo population with spherical distribution condensed first from the primeval protogalactic material. This ancient period of the evolution is called the halo phase of the Galaxy. The halo stars are distributed in a spherical space around the galactic center, with a strong concentration towards it. Their orbits are approximately elongated ellipses, with galactic center in one focus. The rotation of the spherical component (as a whole) around the galactic axis is relatively slow, being about 50 km s^{-1} in the solar vicinity.

From the original spherical shape the protogalactic gas contracted to oblate spheroid which gradually flattened towards the galactic plane, until it ended in a flat disk. The process of contraction from the originally spherical shape to a very flat disk was very fast and lasted only 2×10^8 yr. During the contraction the process of star formation was very active, especially at the beginning, so that the stars of halo and of intermediate populations (in oblate spheroids) represent the major part of stars in our Galaxy.

Simplifying the gravitational contraction of the protogalactic material, we decompose the vector of gravitational attraction of a particle by the rest of protogalaxy into two components: one radial (perpendicular to the axis of rotation) and the other parallel to it, i.e., perpendicular to the galactic plane. The latter force contracted the gas to a flat disk. The radial force increased the angular velocity of revolution of the gas (ω), due to conservation of the angular momentum ($m\omega r^2$, where r is distance from the axis of rotation and m mass of the particle). The contraction (i.e. decrease of r and increase of ω) continued until the gravitational force directed towards the axis was equal to the centrifugal force $m\omega^2 r$. The period of transition from the spherical distribution of the protogalactic primeval gas to the flat-disk distribution of the interstellar gas is called contraction phase of the Galaxy. It was short (2×10^8 yr) and was characterized by a very high birth-rate of stars.

The stars formed during the last phase of the evolution, called disk phase of the Galaxy, are distributed in a very flat disk. They move in nearly circular orbits and their velocity of revolution around the galactic center is about five times larger than for the halo component (i.e., 250 kms^{-1} near the Sun). The disk component of the Galaxy (stellar population I) also contains interstellar gas which is a poor and chemically altered remnant of the original protogalactic gas. Thermonuclear reactions in stars produced heavier elements which were later transferred into the interstellar space and enriched the interstellar gases. The disk phase of the Galaxy has lasted for about 50 galactic yr, i.e., $50 \times (2 \times 10^8$ yr), which is much longer than the duration of the previous contraction phase, viz. one galactic year. The birth-rate of stars during the disk phase, however, has been very low when compared with the contraction phase of the Galaxy. The conditions favorable for formation of large stellar systems – such as giant globular clusters – never occurred during the very long disk phase of the Galaxy.

(c) *Role of Galactic Nuclei in Evolution*

In Section 4.7(d) we discussed activity of galactic nuclei which substantially influences the structure, energy production, and evolution of the galaxies. In the conventional picture of galaxy formation described above, the matter proceeds from rarefied protogalactic gas to compact bodies – mostly stars. The nucleus of galaxies would thus be a secondary phenomenon.

A quite opposite view has been expressed by V. A. Ambartsumian long before importance of galactic nuclei and their activity were known. Galactic nuclei are not a consequence of galaxy formation but they are of primary importance for the life of galaxies and they have existed from the beginning – maybe as extremely dense fragments of the primeval fireball. They are sources of matter and energy for all the phenomena observed in galaxies. According to this theory the evolution should proceed from dense to rarefied; and the observed explosions of galactic nuclei with liberation of immense amounts of energy may represent delayed small scale big bangs of the nuclei. The state of matter in the nuclei is unknown.

To what degree each of these two opposite theories corresponds to actual formation and evolution of galaxies remains still an open question.

5.4. Formation and Evolution of Stars

Stars, once a symbol of invariability, have their fates (*'habent sua fata stellae'*). They are born from interstellar matter by gravitational contraction. Most of their normal life proceeds in converting H to heavier elements by strong and weak interactions. After the final gravitational collapse the star is left without energy sources and it radiates only energy inherited from previous evolution. Electromagnetic interactions are important during the whole life of stars. They regulate energy production by thermonuclear reactions and transport the produced energy from the stars into the interstellar space.

The whole life of a star is characterized by a strong tendency to get rid of the gravitational and nuclear potential energies contained in the protostar; and to come into equilibrium with the cool surrounding universe. In the initial gravitational condensation and in the final gravitational collapse the gravitation draws on the reservoir of potential energy, while during the long thermonuclear period in between the strong interactions liberate most of the energy.

5.4.1. STAR FORMATION FROM INTERSTELLAR MATTER

There is observational evidence that star formation is continuing at present in spiral arms. The birthrate is much slower than in early phases of the Galaxy evolution. The original material has been chemically changed and the birth place has been restricted to spiral arms.

(a) *Interstellar Matter*

The material from which stars are formed has changed considerably since the archaic times of the Protogalaxy: (1) Its amount has been reduced to a few percent since the major part condensed into stars during the contraction phase: (2) The protogalactic H (and He) has beeen partly synthetized into heavier elements by thermonuclear reactions in stars; (3) While the protogalactic gas 10^{10} yr ago had a spherical or spheroidal shape, the interstellar gas is at present concentrated into a flat disk in the galactic plane. There it is compressed by gravity waves into spiral arms. The process of star formation therefore is limited to a relatively small volume of spiral arms, while in contraction phase the stars were born everywhere in the spheroidal protogalaxy; (4) The rotation of the interstellar gas about the axis of the Galaxy is about five times faster than that of the protogalactic gas (250 km s^{-1} compared with 50 km s^{-1} in the solar neighborhood). (5) The evolution of the gas left its traces in the structure of our Galaxy. Spatial distribution, age, chemical composition and dynamical properties of different stellar subsystems (from halo to the extremely flat population – see Table 4.3) reflect the evolution of the gas component.

The interstellar material used for star formation is replenished by plasma ejected from stars (Figure 5.4). The mass balance of matter in the Galaxy is represented in Table 5.1. The turnover rate in both directions of the cycle is several solar masses per year. The total mass of the interstellar matter has been estimated to be 10^9 to $10^{10} M_\odot$. The mean lifetime of an atom in interstellar space would thus be about 10^9 yr before being gravitationally drawn into a protostar.

The kinetic energy in the random flow (6 to 10 km s^{-1}) of interstellar matter and of the cloud motion (a few erg cm^{-3}) is lost by radiation (in collisions or by friction), by twisting of magnetic field lines or it is used by acceleration of cosmic rays by the Fermi mechanism. The explosion of novae and supernovae, expansion of H II regions from bright stars and similar processes supply new kinetic energy to compensate for losses and to keep the energy balance in the interstellar matter.

(b) *Star Formation by Gravitational Contraction*

The interstellar matter is very clumpy with different densities in different places (Table 3.1). Only dark clouds of gas and dust have high enough density and very low temperature to be contracted by self-gravitation and evolve into a protostar. In a dense and cool cloud the gravitation may exceed its internal pressures tending towards expansion of the cloud. The internal pressure is mainly due to heat, radiation, turbulence, magnetic fields and centrifugal forces. For a given mass of the cloud and the internal pressure a minimum density must be reached for the force of contraction to exceed the pressure. For a mass $\sim 10^3 M_\odot$ the density is about 10^{-24} g cm^{-3}; for smaller masses the minimum density is inversely proportional to the square of the cloud mass. For one solar mass it would be $\sim 10^{-18}$ g cm^{-3} or 10^6 atoms per cm^3 and for a mass of 0.01 $M_\odot$ it should be $\sim 10^{-14}$ g cm^{-3}. Such high densities may exist in clouds associated with molecules (Table 3.1).

The stars are formed in groups (associations, galactic clusters). A massive cloud begins to contract even if its density is small, as mentioned above. Thus the gravitational instability affects first a large massive cloud ($10^3 M_\odot$–$10^4 M_\odot$), which breaks into fragments during contraction. If the fragments are able themselves to condense to stars (appropriate mass, low temperature, no turbulence, and sufficient density), then their contraction begins (Figure 5.5). The fragment – probably

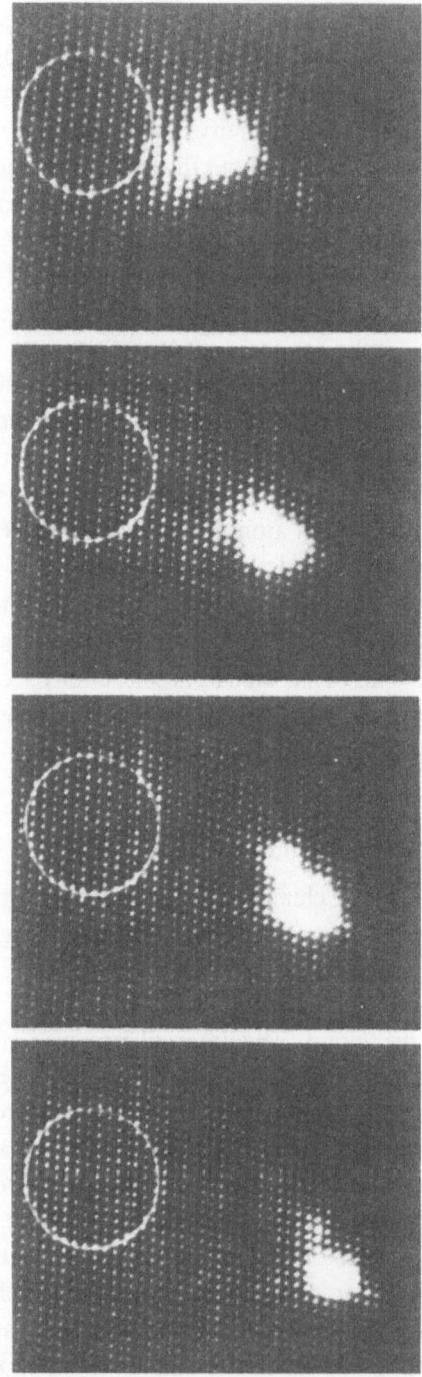

Fig. 5.4. Even relatively quiet stars eject plasma into the interstellar space. On this sequence of radio heliograms on 80 MHz (1969, March 1–2) ejection of material from the Sun may be seen. (CSIRO – Radio Division).

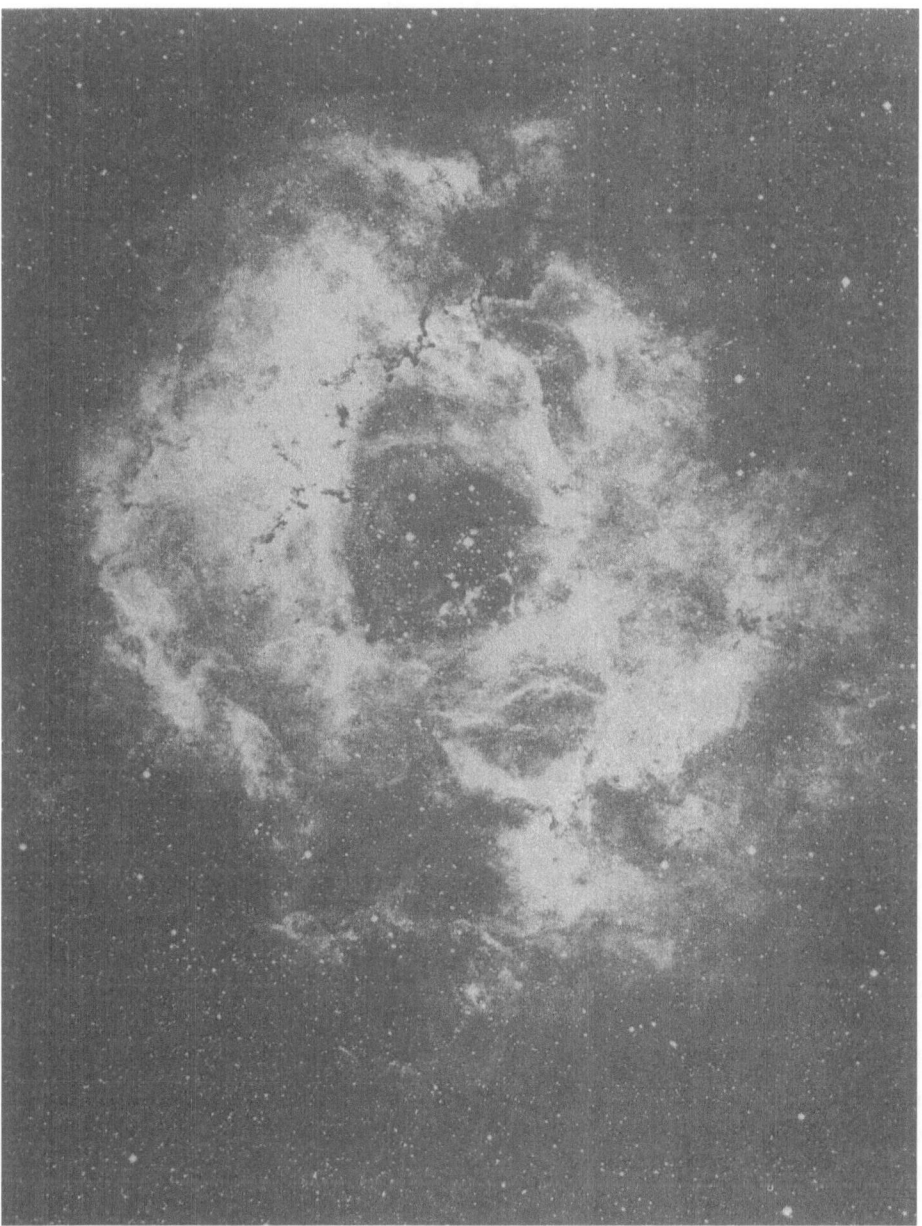

Fig. 5.5.   Rosetta nebula is an example of H II region. In this region in the constellation of Monoceros, about 3500 light years distant, much less than a million years ago groups of newborn stars began to shine. What was left over from the parent cold interstellar gas, from which the stars had been formed, then became ionized and heated by the intense radiation of the newborn stars. Dark interstellar matter can be seen in front of the nebula. It is assumed that protostars condense from such clouds. The mass of the Rosetta nebula is estimated to be 9000 $M_\odot$. As many other H II regions, Rosetta nebula is a radio source (Karl Schwarzschild Observatorium, Tautenburg).

TABLE 5.1

The cycle of matter in the Galaxy

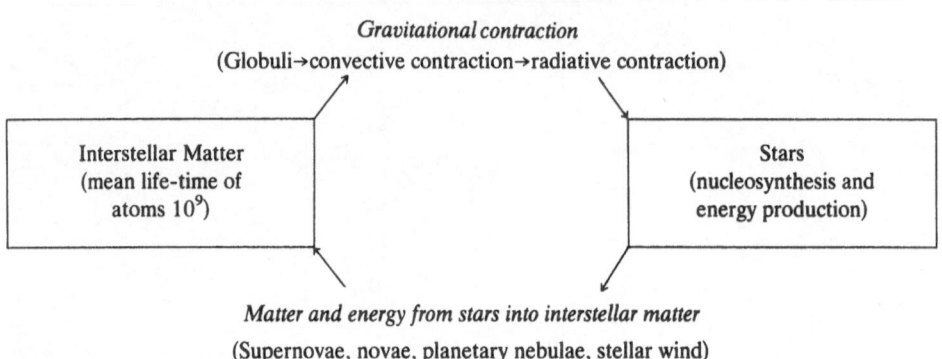

Gravitational contraction

(Globuli→convective contraction→radiative contraction)

| Interstellar Matter (mean life-time of atoms 10^9) | | Stars (nucleosynthesis and energy production) |

Matter and energy from stars into interstellar matter

(Supernovae, novae, planetary nebulae, stellar wind)

observed as a dark globule against bright nebula – will collapse as long as the gravitational forces prevail over the internal pressure. An efficient cooling mechanism will keep the internal pressure low, self-gravitation will dominate, and the collapse will continue.

By gravitational contraction the interior of the protostar is heated. The energy is transported to its surface by convection, owing to the high opacity of the material ($\varkappa \sim \varrho T^{-3.5}$). The whole protostar is convective throughout. Its surface temperature remains nearly constant ($\geqslant 3000$ K) during the contraction. As a result the first part of the evolutionary track in the Hertzsprung-Russell diagram is roughly parallel with the luminosity axis (Hayashi track Figure 5.6).

Later, the central temperature of the protostar increases and opacity decreases to such a value that the transport of liberated (gravitational) energy proceeds by radiation. The luminosity of the protostar stabilizes, the evolutionary track turns to left and becomes roughly parallel with the temperature axis. This period of gravitational contraction is called T Tauri phase. A decrease in the radius increases the central temperature, until thermonuclear reactions can start. The star arrived at the main sequence and will stay there for a long time.

By contraction of the cloud to a protostar and to a main sequence star with radius R, the potential energy

$$E = -G \int_0^M \frac{M_r \, dM_r}{r} \approx \frac{GM^2}{R} \tag{5.14}$$

is liberated. The liberated energy is transformed into heat and into radiation of the protostar:

$$L = -\frac{dE}{dt} \approx -\frac{GM^2}{R^2} \frac{dR}{dt}. \tag{5.15}$$

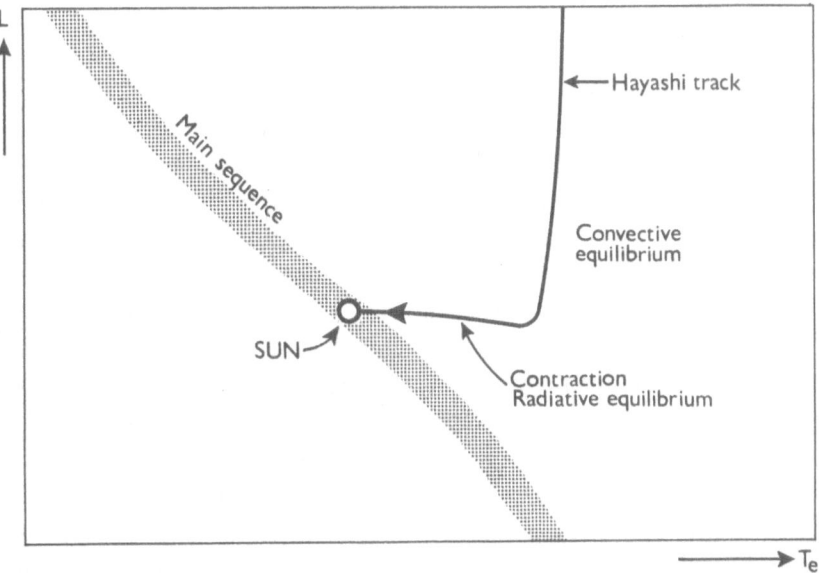

Fig. 5.6.   Contraction of a protostar in the Hertzsprung–Russell diagram. Gravitational potential energy of the extended protostar is the source of its luminosity. Half of the energy is radiated away and the other half heats the interior of the protostar until H begins to burn and the star settles on the main sequence for a long time.

The protostar radiates by decreasing its radius, $dR/dt < 0$ so that the luminosity $L > 0$. From Equation (5.15)

$$dt = -\frac{GM^2}{R^2 L} \, dR \tag{5.16}$$

and for a constant luminosity L

$$t = \frac{GM^2}{RL}. \tag{5.17}$$

The Sun with its present luminosity $L_\odot = 3.9 \times 10^{33} \, \mathrm{erg\,s^{-1}}$ would radiate its gravitational energy in about 10^7 yr.

The rate of evolutionary processes in the protostars depends strongly on their mass. The contraction of the very massive protostars proceeds very fast and is accomplished much earlier (after 10^5 yr) than for protostars with very low mass (10^8 to 10^9 yr). That explains why in the young open clusters (e.g. NGC 2264 in Monoceros) the massive O and B stars are already on the main sequence, while the less massive stars lie above and to the right of it: they had not time enough to reach the main sequence. Among them also are T Tauri stars. Many of the pre-main-sequence protostars appear to be surrounded by gas and dust envelopes – remnants from the parent cloud.

5.4.2. MAIN SEQUENCE PHASE

The gravitational contraction of the protostar comes to a halt when the central temperature reaches values sufficient for hydrogen burning. The main sequence is thus a locus on which gravitational contraction is halted, because the nuclear interactions generate enough energy for balancing the radiation losses of the star. They are also sufficient to produce enough pressure to secure support against gravity.

If there were no exothermic nuclear reactions, the gravitational contraction would supply further energy: half of the gravitational energy would be transformed into thermal energy and the other half remains free for radiation from the stellar surface. That is the case for very light stars ('featherweight stars') with mass less than $0.08 \, M_\odot$. Their central temperature never becomes sufficiently high for H burning (about 7×10^6 K). Evolution of stars with $0.08 \, M_\odot$ was computed to last for 7×10^8 yr before they stabilize on the main sequence. On the other hand, the maximum central temperature for stars with $0.06 \, M_\odot$ will never exceed 2.5×10^6 K and the star cannot settle on the main sequence. Its entire life is dominated exclusively by gravitational contraction, cooling, and a steady decrease of luminosity. The contraction will continue until a fully degenerate black dwarf state is reached. Stars near the lower limit of the main sequence have been observed: Ross 614B at a distance of 13 ly with a mass of $0.062 \, M_\odot$; Wolf 424 at 14 ly with $0.065 \, M_\odot$.

While on the main sequence, a star transforms H into He. The corresponding mass defect 0.006 is transformed into energy, generating about 6×10^{18} erg per 1g of transformed H. Two different reaction sequences participate in the H burning: the proton-proton chain (Table 2.3) which divides into three branches and the CN-cycle which becomes more important for stars with $M > 1.7 \, M_\odot$. The pp-chain is active in less massive stars $M < 1.7 \, M_\odot$ or in any star of pure H (i.e., without C and N atoms). The energy production is sensitive to temperature in the pp-chain:

$$\varepsilon \sim \varrho T^4 \tag{5.18}$$

and very sensitive for the CN-cycle

$$\varepsilon \sim \varrho T^{18}. \tag{5.19}$$

The exponents 4 and 18 are approximate values, real values depend on temperature itself. In the Sun the main energy source is the pp-chain.

The Sun has a mass 2×10^{33} g. The H burning is concentrated to its central core. About 12% of the total H content, viz., 2×10^{32} g may be converted by the Sun during its stay at the main sequence. Thus energy 12×10^{50} erg will be generated during the main-sequence phase of the Sun. Assuming its luminosity is constant $(4 \times 10^{33} \text{ erg s}^{-1})$ the main sequence-phase of the Sun (and any star with the solar mass) should last 3×10^{17} s or 10^{10} yr. Due to the very long duration of H burning, most stars of our Galaxy are now found to be on the main sequence.

A fraction of the energy generated in the main sequence stars is in the form of electron neutrinos (see Table 2.3). The solar neutrinos should represent a detectable flux on the Earth, about 7 SNU; the unit $1 \, SNU = 10^{-36}$ captures per Cl atom of the detecting material (C_2Cl_4) per second. The measured upper limit for the neutrino flux is one order of magnitude lower (i.e., <1 SNU). This large discrepancy remains unexplained. It may be caused by any of the following reasons: (1) the solar model used is wrong; (2) solar neutrinos do not arrive on the Earth; (3) the theoretical probability of neutrino capture is not correct; or (4) the principle of the Ar detection (Equation (2.55)) is wrong (Section 2.3.10).

5.4.3. POST-MAIN-SEQUENCE EVOLUTION

The stars remain on the main sequence or close to it until the H in the core has been converted to He. The burning in the core is replaced by H burning in a shell surrounding the core. The star thus consists of a He core surrounded by a thin energy producing shell and an extended H-rich envelope. When the core contains more than 0.1 of the stellar mass, it becomes unstable and begins to contract fast while the outer envelope expands. Its representative point in the HR diagram moves to the right from the main sequence, i.e., to the red-giant region. The energy during this transition is supplied partly by gravitational contraction of the He core, partly by H burning in a thin shell surrounding the contracting core.

The evolutionary tracks have been computed for stars of different masses (e.g. by Iben) and some are represented in Figure 5.7 and Table 5.2.

As follows from computed models, the densities of the He cores of light stars become so large by the contraction that the core degenerates (Figure 5.8). For heavyweight stars the temperature of the He core remains above the degeneracy temperature (Equation (3.15)) and the hot core behaves like an ideal plasma. A

TABLE 5.2

Evolutionary tracks of stars in the Hertzsprung-Russell diagram. Age for each point in Figure 5.7 is given in millions of years (Models of Icko Iben)

Point	Mass of stars in units $M_{\odot} = 2 \times 10^{33}$ g				
	1	1.5	3.0	5.0	9.0
A	50	18	2.5	0.58	0.15
B	8060	1567	227		21.29
C	9700	1650	239		21.90
D	10236	2036	248		22.08
E	10446	2105	249	70.4	22.13
F	10875	2263	253	70.8	22.14
G			289	78.4	22.73
H			309	85.2	23.15
K			326	87.8	25.74
L					26.23

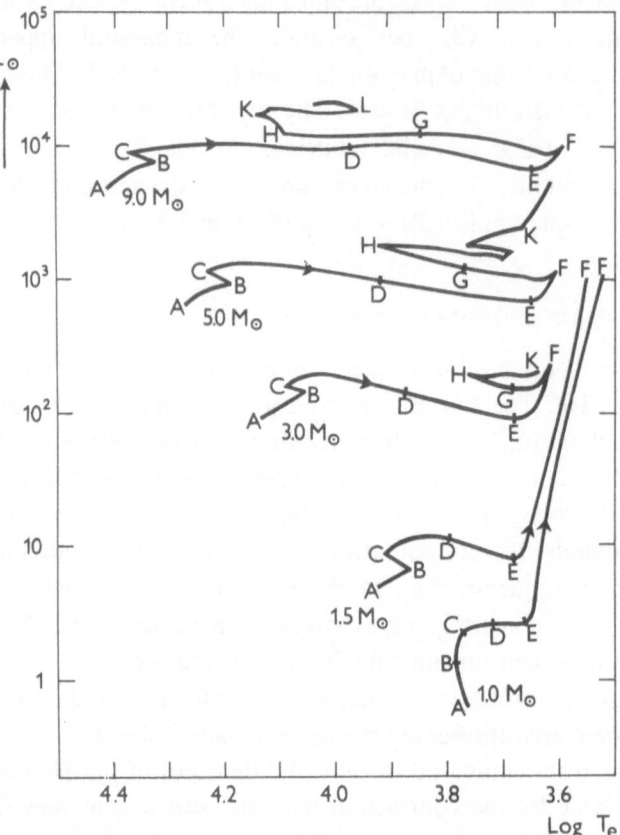

Fig. 5.7. Evolution of stars after the main sequence. The representative points in the Hertzsprung–Russell diagram describe evolutionary tracks. The tracks depend primarily on the stellar mass, as illustrated. The phase A–B corresponds to H burning on the main sequence. The times corresponding to points on the tracks may be seen from the table 5.2.

sequence of thermonuclear reactions generates energy and produces heavier elements. Where the sequence stops depends mainly on the mass of the star (Table 5.3).

(a) *Helium Burning*

At temperatures of 10^8 K and densities of 10^4 g cm^{-3}, He itself becomes a fuel. The He burning is dominated by two reactions: triple-alpha reaction which produces C

$$3\,^4\text{He} \rightarrow {}^{12}\text{C} + \gamma \tag{5.20}$$

and the α-particle capture

$$^{12}\text{C} + {}^4\text{He} \rightarrow {}^{16}\text{O} + \gamma \tag{5.21}$$

which destroys it. A further step

$$^{16}\text{O} + {}^4\text{He} \rightarrow {}^{20}\text{Ne} + \gamma \tag{5.22}$$

does not seem important, due to the lack of He.

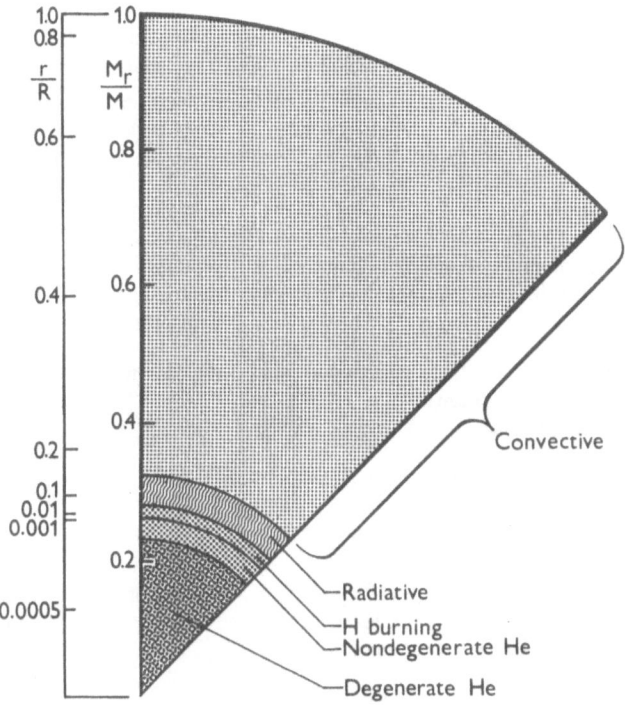

Fig. 5.8. Model of a red giant. The mass of the star is 1.3 $M_\odot$ and a quarter of its mass has already been converted to He. The major part of the He core is degenerate.

(b) *Carbon Burning*

The next major step in the evolution of middleweight and heavyweight stars is the C burning. The stars may burn C in the red-giant branch at temperatures about 0.8×10^9 K. Products of the C burning are nuceli near Mg, e.g.

$$^{12}C + {}^{12}C \begin{cases} {}^{20}Ne + {}^4He \\ {}^{23}Na + {}^1H \\ {}^{23}Mg + n \\ {}^{24}Mg + \gamma. \end{cases} \qquad (5.23)$$

Lightweight stars cannot reach the C burning stage at all. Middleweight stars which have burnt out their He to C in their degenerate core (see Table 5.3), may also synthetize Mg elements from the C nuclei; but in the middleweight stars the C burning is explosive and leads to a supernova explosion. The explosive C burning in supernova explosions (at temperatures 3×10^9 K to 7×10^9 K) will form most nuclei between ^{20}Ne and ^{62}Ni.

TABLE 5.3

Post-main-sequence evolution of stars. The table should serve only for a first orientation. Not much confidence should be put into the mass limits. Caution is recommended in using the table

Star class	Heavy weight	Middle weight	Light weight	Feather weight
initial mass (limits)	$>8\,M_\odot$	$8\,M_\odot\text{–}4\,M_\odot$	$4\,M_\odot\text{–}0.08\,M_\odot$	$<0.08\,M_\odot$
energy loss	photons for H and He burning neutrinos for C burning and beyond		photons neutrinos negligible	photons no neutrinos
stellar core	non-degenerate high temperature ideal plasma	degenerate core	degenerate	degenerate
central temperature	up to $3\text{–}5\times10^9\,\text{K}$ in equilibrium	$3\text{–}7\times10^9\,\text{K}$ during detonation	$\sim10^8\,\text{K}$	$<10^7\,\text{K}$
Sources of stellar luminosity	burning of: H, He, C (non-explosive), O, Si; all fusion reactions releasing energy	burning of: H, He, C (explosive)	burning of: H, He	gravitational contraction; no stay on main sequence
mass loss	?	substantial to $\leqslant 2\,M_\odot$	large to $\leqslant 1.4\,M_\odot$	unimportant
destruction	gravitational collapse (supernova?)	supernova explosion	mass ejection (planetary nebula)	cooling
end product	black hole	neutron star (pulsar)	white dwarf	infrared dwarf
pressure against gravitation	none whatever	neutron degeneracy	electron degeneracy	electron degeneracy

(c) Oxygen Burning

Because the charge of the O nucleus is relatively high, a considerable increase in temperature is required before further reactions in the sequence of thermonuclear reactions can occur. When the temperature rises to $T \lesssim 2\times10^9\,\text{K}$ the O nuclei are synthetized to nuclei in the vicinity of Si, such as

$$^{16}\text{O}+{}^{16}\text{O}\begin{cases} {}^{28}\text{Si}+{}^4\text{He} \\ {}^{31}\text{P}+\text{p} \\ {}^{31}\text{S}+\text{n} \\ {}^{32}\text{S}+\gamma. \end{cases} \qquad (5.24)$$

The O burning may occur under explosive conditions, similarly to explosive C burning. It is interesting to notice that the explosive C and O burning explains quite well the solar-system abundances of the nuclides in the mass range $20 \le A \le 40$, while the hydrostatic burning in heavy weight stars cannot explain them.

(d) *Silicon Burning*

The next major step in the sequence of thermonuclear reactions in massive stars is the Si burning. The Si nuclei ^{28}Si produced in O burning (Equation (5.24)) have a high nuclear charge and there is large coulomb repulsion between them. Direct nuclear interaction analogous to (5.20), (5.23) and (5.24), i.e. ^{28}Si + ^{28}Si $\rightarrow$ ^{56}Ni + γ, is not possible even at temperatures above 2×10^9 K. The heavier elements are produced from Si in a different manner. For temperatures $T > 2 \times 10^9$ K or mean photon energy larger than 2×10^5 eV (1 eV is associated with temperatures 1.6×10^{-12} erg k^{-1} = 11 600 K). Due to their Planckian distribution, there are many photons with much higher energies sufficient to release from the nuclei synthetized by C and O burning free neutrons, protons and α-particles. Symbolically,

$$^{28}\text{Si} + \text{several } \gamma\text{-photons} \rightarrow 7\,^4\text{He}. \tag{5.25}$$

The released α-particles and other fragments are recaptured by ^{28}Si and by some heavier nuclei. These processes are called Si burning, although nuclei of other nuclei are also involved. The binding energy of nucleons increases until the Fe group (Figure 2.1) so that the processes of Si burning are exothermic and contribute to the star luminosity. The whole sequence of thermonuclear reactions from pp-chain to Si burning gradually increased the binding energy of the nucleons. The free protons in the core of a protostar have been gradually bound during the stellar evolution to compact nuclei of the Fe group; they lost 0.9 MeV of its rest mass and thus contributed to the luminosity of stars on the main sequence and during the post-main-sequence evolution.

(e) *Equilibrium Processes*

Under the conditions in the core of heavyweight stars (Figure 5.9), namely $T \sim 3 \times 10^9$ K to 5×10^9 K, the nuclear reactions become very frequent and a state of nuclear equilibrium is approached. A complicated set of reactions (not all of the reaction cross sections are known) tends to squeeze out the last remnants of nuclear energy in the core material: it tends to form an abundance peak around the Fe (Figure 4.2). The elements of the Fe group have the highest binding energy per nucleon (Figure 2.1), so that no further nuclear energy is available from them any more. The thermonuclear evolution of massive stars which started with H comes to its end with the Fe group elements in their core. No more nuclear energy can be squeezed out from the hot Fe-rich plasma. Gravitation remains the only source for the final destruction of the stars (gravitational collapse).

Only the major stages of energy and element formation by strong interactions have been mentioned. Before discussing the destruction of stars by gravitation, we should

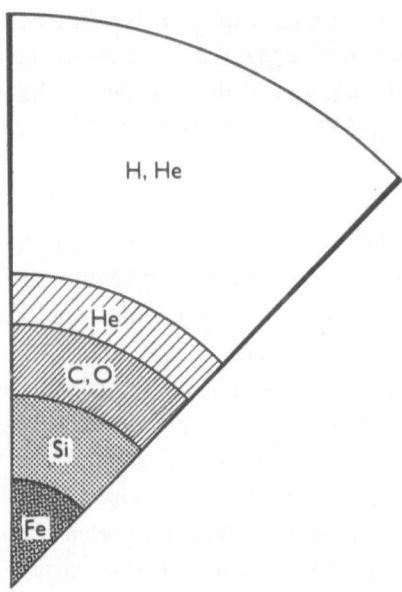

Fig. 5.9.   A heavyweight star at the end of thermonuclear evolution and before the gravitational collapse.
The whole scale of thermonuclear reactions is active at different depths in the interior.

mention one nuclear process in thermonuclear evolution, which is more important for the nucleosynthesis of elements with atomic mass $A > 60$ than for energy generation, namely the slow neutron capture.

(f) Slow Neutron-Capture

For building up nuclei heavier than Fe group nuclei, the neutron capture plays a fundamental role. The new nucleus produced by a neutron capture, i.e.

$$(Z, A) + n \to (Z, A + 1),$$

e.g.

$$^{58}Fe + n \to {}^{59}Fe,$$

(5.26)

is often followed by a beta decay, such that

$$(Z, A + 1) \to (Z + 1, A + 1) + e + \tilde{\nu}_e,$$

e.g.

$$^{59}Fe \to {}^{59}Co + e + \tilde{\nu}_e.$$

(5.27)

The beta decay lifetime for the nuclide ^{59}Fe is 45 days.

The next neutron may be captured either immediately after Equation (5.26), i.e., before Equation (5.27) can take place:

$$(Z, A + 1) + n \to (Z, A + 2).$$

(5.28)

The processes (5.26) and (5.28) are called r-process and they may occur only in rich fluxes of neutrons, such as exist in supernovae ejecta (Section 5.4.4(c)).

During the thermonuclear period neutrons are produced in stars in small quantities, so that there is enough time for a beta-decay between two successive neutron captures. The neutron capture on a slow time scale is called slow neutron capture or s-process.

The build-up of nuclides by slow neutron capture continues up to mass number $A = 210$. The process is slow and the beta decay constrains the built-up nuclei to the bottom of the stability valley (Figure 2.2). There exists observational evidence that the s-process has been active in some stars: technetium lines in some red giants, abundance of Hg isotopes, excess of Ba and rare earths etc. Technetium (Tc, $Z = 43$) is an element with no stable isotopes; their lifetimes range from seconds to 2.6×10^6 yr. Even the longest lifetime is much shorter than the lifetime of stars. That means that Tc is currently being produced in stars in the spectra of which it is observed. Its production proceeds by s-process.

(g) Explosive Thermonuclear Reactions

The sequence of thermonuclear reactions develops with increasing temperature. A minimum temperature is needed for each of the reactions (from pp-chain to Si burning) and the reactions start as soon as the minimum temperature is reached by gravitational contraction of the core.

In Section 5.4.3(c) explosive C burning and O burning has been mentioned. Such processes, and in general all explosive thermonuclear reactions, occur when the stellar material is brought very rapidly to a much higher temperature than the minimum temperature needed. As the thermonuclear reaction rate depends very strongly on temperature (for example formulae (5.18) and (5.19) for H), the energy generation of the suddenly overheated stellar plasma has an explosive character and the abundances of produced elements differ from those produced in the quiet quasi-static evolution.

5.4.4. DESTRUCTION OF STARS

The evolution, destruction and end product of a star depends on its mass. In a simplified form the whole history of stars is represented in Table 5.3. All the end products consist of matter at high density. The compression from about 10^{-20} g cm^{-3} (in the interstellar cloud) to 10^{10} g cm^{-3} or more (in neutron stars) has been done during the stellar evolution by self-gravitation. The gravitational potential energy is an important source of stellar luminosity and for the featherweight stars it is the only one.

The gravitational forces of the end products of stellar evolution – especially in the case of neutron stars and black holes – are so intense that relativistic effects become important for their structure and behavior. The deflection of star light passing close to (1) the Sun is 1.7″ or $\sim 10^{-3}$ deg (Section 2.4.4(a)); (2) a white dwarf is about

0.1 deg; (3) a neutron star may be from 10 to 100 deg; (4) a black hole well; there is complete absorption of the light photons which fall into the Schwarzschild sphere.

(a) *Infrared and Black Dwarfs*

The featherweight stars with mass $M < 0.08 \, M_\odot$ do not have their plasma treated by thermonuclear reactions. They do not reach the minimum temperature for H burning, do not stop at the main sequence and the notion of 'main sequence' for featherweight stars has no sense. The stars live from gravitation only; they contract, degenerate and cool. Their very faint luminosity in the IR makes them difficult to observe. They should, however, exist because there is no reason why the fragmentation of the large parent cloud (Section 5.4.1(b)) should work only for masses $M > 0.08 \, M_\odot$ giving rise to the main sequence stars.

(b) *White Dwarfs*

The white dwarfs (Section 4.5.2) are the end product of lightweight stars that have never reached the stage of C-burning. There is observational evidence suggesting that mass ejection from lightweight stars plays a decisive role during their red-giant phase. The upper mass limit for white dwarfs is $1.4 \, M_\odot$ so that a heavier star (e.g., 2 or $3 \, M_\odot$) must get rid of the surplus mass to become a white dwarf. The red giants are described as white dwarfs surrounded by an extensive envelope. The envelope may be driven away by radiation pressure and runaway pulsations. Its large portion may appear as a planetary nebula. This class of nebulae which has nothing in common with the planets – except for their appearance in telescope – consists of a hot star (nucleus) and an envelope ejected probably in a non-explosive manner. The nucleus is probably a star which has exhausted all its thermonuclear possibilities and is not yet strongly degenerate. It contracts, increases its degeneracy, and cools to become a white dwarf.

(c) *Supernovae, Neutron Stars, and Pulsars*

Neutron stars are produced during supernova explosion of middleweight stars (Section 4.5.3 and Table 5.3). The C burning in a highly degenerate core may be explosive and followed by a complete violent explosion of the whole star; in this case the thermonuclear burning would run fast all the way to the Fe elements peak, and matter is hurled into the interstellar space with no remnant left over.

If, however, there is an efficient neutrino cooling (Section 2.3.11), the C burning is slowed down and the core of the star falls towards its center (the so-called implosion of the stellar core), because there is not a sufficient pressure to resist the self-gravitation. The implosion of the core may be slowed down by the pressure of the degenerate electron gas. The implosion is halted when nuclear densities $\geqslant 10^{14} \, \mathrm{g \, cm^{-3}}$ are reached; for at that stage the degeneracy pressure of the neutron gas is sufficient to stop the imploding masses.

When the infalling masses of the envelope hit the neutron core of the star (which has just been formed) a strong compression wave (shock wave, detonation wave) develops, propagating outwards and heating the envelope to such temperatures that many thermonuclear reactions flare up in the unprocessed plasma. In temperatures of the order of 10^{11} K (or 10^7 eV) nuclei are fragmented and abundant fluxes of neutrons and protons are thus generated.

In the rich neutron fluxes the interval between two neutron captures is shorter than the β-decay time (Equations (5.26) and (5.28)). The process is called rapid neutron capture, neutron capture on a rapid time scale, or simply r-process. Several or many neutrons may be captured by a nucleus exposed to a rich neutron flux before the nucleus becomes unstable and decays towards the bottom of the stability valley (Figure 2.3). By the r-process unstable heavy nuclides are formed on the neutron-rich slope of the chart of nuclides. By β-decay they shift towards the bottom of the valley, giving rise to more stable superheavy nuclei.

The envelope overheated by the compression wave and subsequent thermonuclear reactions thus explodes synthetizing many heavy and superheavy nuclei and liberating huge amounts of energy. This complex event at the end of thermonuclear evolution of middleweight stars (and maybe also of heavyweight stars) is called a supernova explosion. The amounts of energy liberated within hours and days ($\sim 10^{51}$ erg) are comparable with the energy radiated by our Sun during its whole life on the main sequence (10^{10} yr). Although the supernova explosion represents only an extremely short fraction of the lifetime of the stars, a substantial fraction of the heavy nuclei and superheavy nuclei observed in nature has resulted from such energetic events. Moreover, the violent supernova explosions have dispersed the envelope, and may also be a part of the core, into the interstellar space (Figure 5.10). The interstellar matter is thus enriched with heavy and superheavy elements. Part of the liberated energy appears under the form of cosmic radiation; the supernovae are an important source of the relativistic particles.

Pulsars are considered to be the neutron stars – i.e., remnants of the core of the middleweight stars after a supernova explosion. The narrowness of the pulse (~ 20 ms) implies that the source is much smaller than any normal plasma star (compare the Section 4.5.3(c)). The signals are repeated with a high regularity (periods range from 0.03 to 4 s) which can be explained by a rapid rotation of the neutron star. The pulsar period however, is observed to be very slowly increasing. The gradual lengthening of the period is interpreted as dissipation of rotational energy of the neutron star into the surrounding interstellar medium. It also implies that the pulsars with the shortest period are the youngest neutron stars (e.g., the Crab pulsar dates from A.D. 1054 when the supernova was historically recorded).

Calculations of the explosive processes show that the C burning in a highly degenerate core of middleweight stars may lead to their complete dispersal, without any neutron star left over. At the present, one does not exactly know whether both processes are acting in destructing middleweight stars. It is worth mentioning that in some supernova remnants no pulsars have been found (e.g. the 1572 and 1604 supernovae, both belonging to the Population II): either no neutron star has been left over, or its axis of rotation has an inconvenient orientation towards us (Section

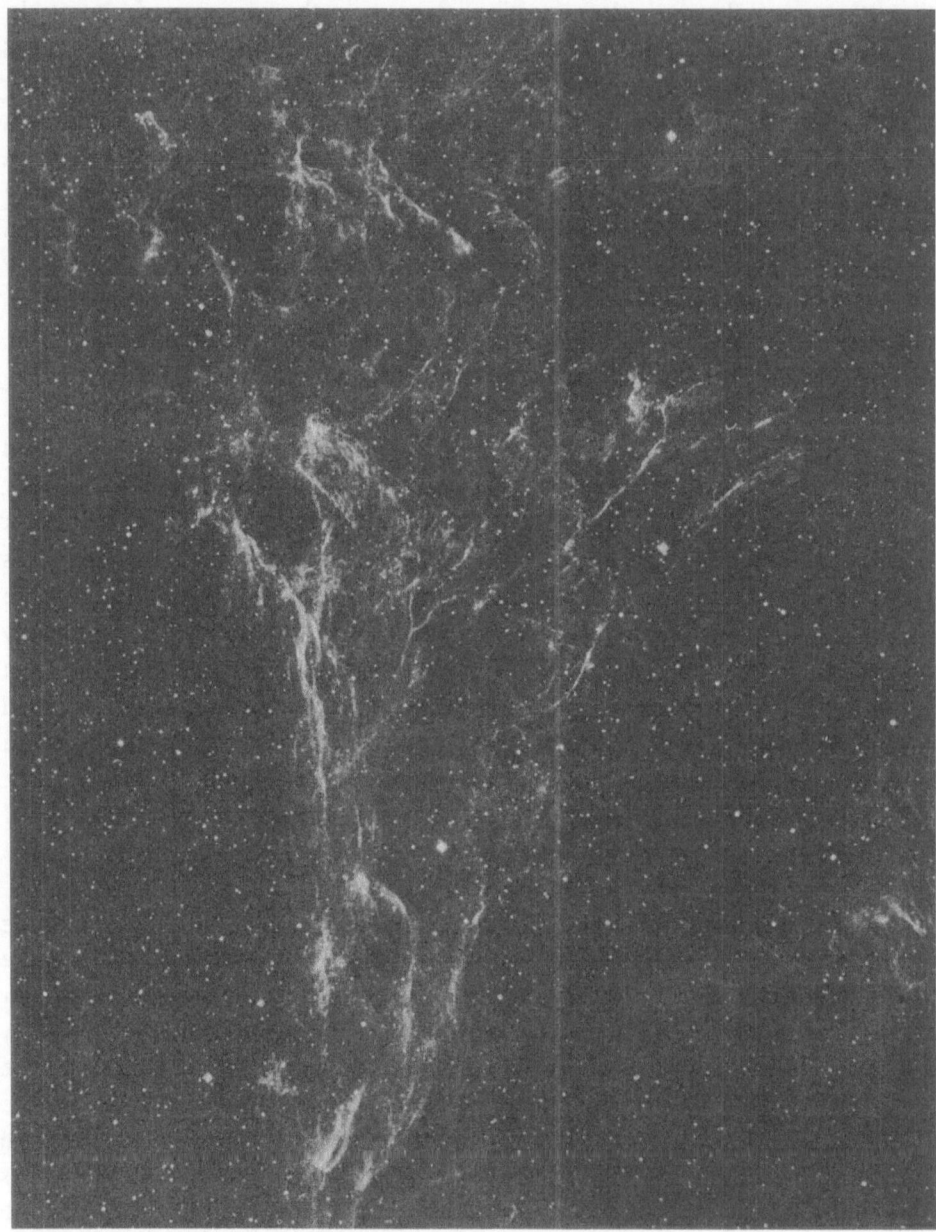

Fig. 5.10.   Some remnants of an ancient supernova (Cygnus Loop, Veil Nebula, Network Nebula). The original expansion velocities (10^3 to 10^4 km s^{-1}) have been slowed down to a hundred km s^{-1}. Ultimately the remnants will slow down to velocities of interstellar matter (about 10 km s^{-1}). The total mass of all the nebula is about 2 $M_\odot$. It is a source of radio and X-ray radiation, roughly 2000 light years distant from us (Karl Schwarzschild Observatorium, Tautenburg).

4.5.3). In any case, the supernova explosions lead to liberation of huge amounts of energy in a short time, to the synthesis of most of the heavy nuclides, to the ejection of the major part (if not of all) of the stellar material and to an acceleration of plasma ions and electrons to cosmic-ray energies.

(d) *Black Holes*

As may be seen from Table 5.3, the very massive stars with initial mass $M > 8 \, M_\odot$. have a non-degenerate core and survive the C-burning, which is non-explosive. After the Fe peak elements have been formed and the core cooled by neutrino emission (Section 2.3.11) the core (and eventually the envelope too) are squeezed by gravitation towards the star's center, below the gravitational radius. If the gravitational attraction between elementary particles does not alter its character for extremely short distances and if the velocity of light represents a limit for the energy transport, then there is no force which could stop the implosion. The matter of the core pours torrentially inward from all directions to occupy an ever smaller volume. It may be slowed down by neutron degeneracy pressure and by repulsive forces of the baryons (when they approach < 1 fermi from each other), but afterwards the speed picks up again.

It is not known how much of the envelope material is swallowed by the collapsing core, how much of it is ejected into the interstellar space and whether a supernova event is observed or not. What is certain, however, is that once the collapsing matter gets below the gravitational radius, no photon and no other particle can escape from it.

The sequence of events during a (spherical) collapse follows from metrics and has been described many times (Figure 5.11). The time scale on the collapsing star is totally different from the time-scale far outside due to difference in gravitational potential. The observer on the collapsing star transmits time signals at some wavelength (by his standard) and at equal intervals by his standard clock. The distant observer first receives equally spaced signals. As the collapse proceeds and radius of the star approaches the Schwarzschild gravitational radius (Equation (2.67)), the signals arrive to the distant observer in longer intervals and on longer wavelengths. After the surface of the collapsing star has passed through the Schwarzschild radius, no signals can leave it any more. The observer on the star crosses the Schwarzschild radius smoothly and after a finite time (counted in seconds by his clock). His density is predicted to grow to infinity in a very short time. He can never retrace his path out of the black hole. For the distant observer the collapse lasts infinitely and the signal frequency decreases to zero.

Properties of matter in this final stage of massive stars are not known. All particularities of stellar plasma disappear during the collapse. What then remains? The mass which binds the planets by gravitation as before, electric charge (if the star as a whole was charged) and angular momentum (if the star rotated before its collapse). The black hole might be characterized by mass, electric charge and angular

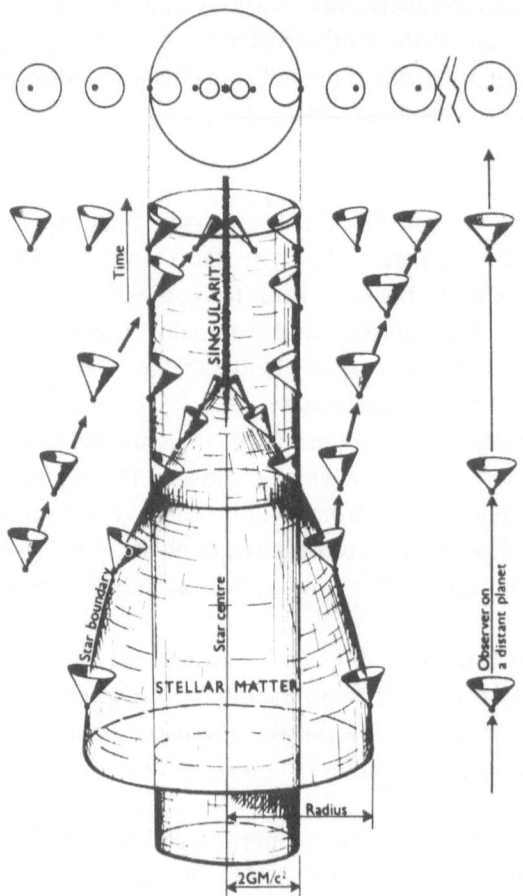

Fig. 5.11. Gravitational collapse of a spherical star. Time goes upwards and one spatial coordinate is suppressed. The history of the star's boundary and the star's center is indicated. The histories of photons emitted at various events (fat dots) are represented by 'the light cones'. As the collapse proceeds, light is more and more dragged inwards due to strong gravitational field. The cylindrical surface (above the cross section with the collapsing star's surface) shows the horizon. It represents the histories of photons which neither escape to infinity nor fall to the singularity. (The lower part of the cylinder has no physical meaning, it indicates gravitational radius of the stellar mass.)

momentum. All the other properties (e.g. quantum properties of elementary particles) are lost during the collapse. It would be impossible, for example, to distinguish between a black hole of ordinary matter and one of antimatter.

Photons and particles from outside are drawn down into the black hole only to increase its mass, so that gravitational attraction eventually changes its charge and angular momentum. The gravitational attraction represents the only hope to discover the black holes. Their complete blackness makes a direct search for isolated black holes hopeless. Gravitational effects external to the black hole (such as black hole with a normal star in a binary system, gravitational waves generated by a non-spherical collapse, contribution to the overall mass density of the universe) seem at present to be the only way of observing them (Section 4.5.4).

5.5. Origin and Evolution of the Solar System

"We enter here in one of the most obscure chapters of our book" (H. Reeves in *L'origine du système solaire*, CNRS 1972, p. 41). Much effort has been expended to explain the origin and evolution of the solar system. It is our home in the universe and there is a lot of observed facts on which a solid theory of its origin and evolution can be built. Different theories have been proposed but it seems at present that the Kant–Laplace theory is the most capable of explaining the observed facts.

Kant was the first who treated the origin according to Newton's laws ('nach Newtonschen Grundsätzen') in *Allgemeine Naturgeschichte und Theorie des Himmels*. The modern version of the theory also explains those properties of the solar system which do not follow from the laws of the celestial mechanics.

5.5.1. PRIMITIVE NEBULA

The primitive nebula (called also presolar nebula, solar nebula, protoplanetary nebula, primeval nebula, parent nebula) is the common ancestor of the Sun and its planetary system. It originated as a fragment of a large ($\sim 10^3 M_\odot$) interstellar cloud of gas and dust – presumably similar to the Orion nebula (Section 5.4.1). The collapse of the large cloud (~ 10 ly) started when it drifted across one of the spiral arms of the Galaxy.

The fragmentation of the large cloud into a system of smaller and denser clouds was described in the chapter on star formation. It is most likely that our Sun with its planetary system was formed along with $\sim 10^3$ stars as a member of a galactic cluster or association. Since then the cluster of sister-stars has been dispersed, so that all information on the original large parent interstellar cloud is lost.

There is no consensus on the collapse of the primitive nebula (one of the fragments), on its mass, size, temperature, pressure, internal motions, chemical and physical processes. Its final size after contraction might have been about $10^4 R_\odot$, about the diameter of the orbit of Pluto. In consequence of the rotation, the primitive nebula was flattened to a disk. In its central part the Protosun condensed ($\sim 5.5 \times 10^9$ yr ago). The planets, their satellites, asteroids, comets and the meteoritic complex formed later from the nebular disk rotating around the Protosun.

(a) *Condensation Sequence*

It is assumed that the protoplanetary nebula (i.e., the nebular disk) passed through a high temperature stage. The protoplanetary nebula was heated by its gravitational contraction; the nebula was opaque so that the liberated infrared radiation could not escape from its central parts. High temperature was therefore in the central regions, close to the luminous Protosun, while at the outer edges the nebular disk remained cold (a few tens of degrees K). As the nebula cooled, various minerals began to condense (Figure 5.12). Some highly refractory compounds of Ca, Al, Mg and Ti appeared first, below 2000 K (CaO, Al$_2$O$_3$, rare-earth oxides, CaTiO$_3$, CaSiO$_3$, MgAl$_2$O$_4$, ...). Other refractory elements from the 2nd to 5th columns of the periodic table (plus Al and Si) followed. When the nebula cooled to $T < 1500$ K a metallic Fe–Ni alloy (similar to that found in meteorites) and magnesium silicates condensed.

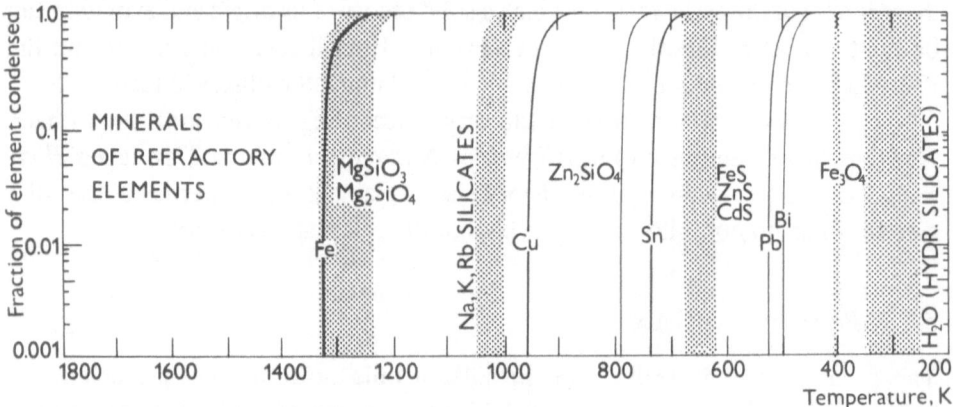

Fig. 5.12. Condensation sequence of the cooling protoplanetary nebula. The first materials to condense were a group of rare minerals with a high content of refractory elements (Ca, Mg, Al, Ti). The first major concentrate were Fe and Mg silicates (olivine, pyroxene). The sequence from left to the right corresponds to decreasing temperature (and increasing time) (after Larimer).

Then at 1100 to 1000 K alkali silicates and alkali metal oxides (as Na$_2$O) condensed and a sequence of many different chemical reactions took place with the decreasing temperature in the protoplanetary nebula (e.g., reactions of Na with aluminium oxide and silicates to produce feldspar and similar minerals). To this temperature about 90% of the material in chondritic meteorites condensed. Only H, C, N, O, S and some trace elements remained in the gaseous state.

At about 700 K S began to condense on solid Fe grains

$$Fe + H_2S \rightleftarrows FeS + H_2 \qquad\qquad (5.29)$$

and produced the mineral troilite FeS.

At 400 K the remaining Fe oxidized to minerals (e.g., olivine) and the water vapor combined with minerals (e.g., with olivine to produce serpentine). Finally at 170 K water vapor condensed into ice.

Ammonia condensed into solid hydrate NH$_3$.H$_2$O at ~ 150 K. At still lower temperatures (50 to 100 K) another solid hydrate condensed, viz. CH$_4$.7H$_2$O. Titan, the largest Saturn satellite with an NH$_4$-rich atmosphere, has probably a solid crust of this hydrate; by melting the methane gas is liberated to form the atmosphere. At last solid methane grains formed at temperatures 20 to 50 K. The density of solid NH$_4$ is low (about $0.6 \, \text{g cm}^{-3}$) and some small satellites of Saturn, Uranus and Neptune seem to have comparable densities.

Densities and chemical composition of the planetary system are explained as a direct consequence of the variation of temperature in the protoplanetary disk. Abundances of elements and of molecules (e.g., volatile metals content) may thus serve as a cosmo-thermometer to infer accretion temperatures of the parent bodies of planets and meteorite parent bodies. Thus for example, the bulk of the accretion between 1 and 3 A.U. took place at a temperature about 450 K and the entire temperature drop from the beginning to the end of the accretion was $\leqslant 80$ K. Another example of accretion-temperature estimates are carbonaceous chondrites which accreted at about 360 K. Organic molecules were formed from reactions of CO and H_2 (with hydrated silicates and Fe_3O_4 acting as catalysts). Generally speaking, the chemical composition must reflect above all the temperature conditions in the protoplanetary nebula whereas the dependence on pressure is weak.

(b) *Gravitational Focusing*

After the solid grains of various minerals condensed out of the protoplanetary nebula, they settled towards the midplane of the disk. A much thinner disk of solid grains and pebbles should have been formed within the thick disk of gases. The differentiation was accomplished in a very short time, probably during a few years. A similar settling of solid grains into a thin disk (gravitational focusing) may be seen in Saturn rings. For several thousand years the gas component of the protoplanetary nebula and its thin solid component pursued different evolution.

(c) *Fragmentation of the Disk*

Instabilities broke up the disk of solid particles into fragments which then accreted by self-gravitation, if their mass had been sufficiently large. The accretion led to the formation of large bodies – protoasteroids, satellites, terrestrial planets and rocky cores that would later become giant planets.

(d) *Accretion and Sweeping Out of Gases*

The gases in the protoplanetary nebula were either accreted on the rocky cores to form the giant planets or they were blown away by intense corpuscular radiation of the Protosun. The Protosun in a complete convection had an extended corona, intense solar wind and was apparently very active. It ejected plasma together with frozen-in magnetic fields. All the material, viz. that ejected from the Protosun and remnants of gas and dust grains from the protoplanetary nebula, was transported beyond the limits of the planetary system, in the near interstellar space. The Oort's cloud of comets might have been formed from that material.

(e) *Differentiation of Large Bodies*

The accumulation of solid material produced homogeneous bodies. The bodies larger than a few hundred km probably melted by heat from radioactive elements and acquired a differentiated structure (core-mantle-crust-atmosphere). The internal structure of our Earth shows that it is a highly differentiated body.

The interior of our Earth was heated by two sources: (1) gravitational energy liberated during the accumulation of the planet, and (2) radioactive decay which produced fifteen times more heat per second than it does today. The increasing temperature caused progressive melting of the solid material in the interior. The molten material was differentiated by the gravitational field: heavy elements (like Fe and Ni dropped towards the Earth's center while the lighter constituents (e.g. silicates) were brought towards the surface by buoyancy forces. The core and mantle of the Earth were thus formed. Later the mantle solidified, the heat transport by convection stopped and it became an insulating envelope for the core. With solidification of the mantle ($\sim 4 \times 10^9$ yr ago) begins the geological history of the Earth.

Also the Moon underwent differentiation in its early history. It had been accreted from solid particles in the circumterrestrial space. The accumulation of both the Earth and the Moon might have been simultaneous. (There are serious objections to the other two theories of the Moon origin – viz., that it has been captured by the Earth as a finished body or that it has been fissioned from the Earth).

However, the Moon no longer moves in the orbit in which it was accumulated. The tides transfer angular momentum from the Earth's rotation to the lunar orbit. As a consequence both the distance of the Moon and the length of the terrestrial day increase. The history of the Moon's orbit has been calculated and the results indicate, that the Moon should have been accumulated somewhere between one fifth and one half of its present distance from the Earth.

5.5.2. Evolution of planetary atmospheres

The present planetary atmospheres have a wide variety of chemical composition (Section 4.4.5(d), 4.3.1 (c), (d), (e)), although the planets were formed out of the same protoplanetary nebula. Only Jupiter and Saturn retained the original composition of the nebula. If the terrestrial planets had retained the original composition (Figure 4.2), the Protoearth, Protomars and Protovenus would have been about six hundred times more massive than they are and they would be essentially composed of gaseous H_2 and He. Also the content of other inert gases (Ne, Ar, Kr, Xe) would be by as much as 10^6 higher and they would certainly not be 'rare' gases in our atmospheres.

5.5.3. Evolution of the earth

The chemical composition of our planet is characterized by a severe deficiency of volatile elements (H, He, Ne, Ar, Kr, C, N), relatively to the nonvolatile ones (such as Mg, Si, Al). The volatile elements were either never accreted or, if they were, the Earth lost them at temperatures of several hundred degrees. It is therefore probable that the Earth was devoid of its extended gaseous envelope in its earliest history. Its present atmosphere is a secondary phenomenon. There exists convincing geochemical evidence that the atmosphere and hydrosphere were formed during geological history by several processes:

(1) Gases were released from the solid globe by volcanoes and during the crystallization of magma. These processes produced mainly H_2O and CO_2, small amounts of N_2, some H_2, SO_2, H_2S, HCl and HF. The volcanic activity was the most important source of CO_2 and N_2 for the atmosphere and it supplied all the H_2O for the hydrosphere. No free O_2 was contained in volcanic exhalations, since any O_2 was bound to compounds, such as oxides, silicates or water.

(2) The atmosphere of CO_2 and N_2 was transparent even for UV radiation. Photodissociation of the H_2O produced free O_2, which in turn absorbed the UV radiation and protected the lower layers of the atmosphere. These two competing processes – i.e. photoionization and absorption – could generate only a small amount of free O_2 in the primitive atmosphere, about 0.1% of the present amount. The autoregulating process (called Urey effect) is caused by the fact that the absorption by O_2 and dissociation of H_2O occur in the same wavelength range. In the prebiological atmosphere the photodissociation of H_2O was the only source of O_2.

(3) Photosynthesis of green plants became the major source of atmospheric O_2. Thus the evolution of the terrestrial atmosphere was decisively influenced by the origin of life and the evolution of life itself was dependent on the atmosphere: (a) the atmosphere provided raw material for synthesis of organic molecules; (b) it protected living organisms against harmful radiation from cosmic space, and (c) it maintained convenient temperature conditions by greenhouse effect and atmospheric circulation.

The high abundance of O_2 (21%) in our atmosphere is unique in the whole planetary system. While most living organisms depend on O_2 and its high concentration, the life could not have originated in presence of such a high amount of O_2. The organic molecules characteristic for living matter (particularly proteins and nucleic acids) oxidize fast in the presence of O_2. At present, they can therefore be formed from inorganic substances only by living organisms. In the prebiological times they could be formed only in an O_2-free atmosphere – consisting of CO_2, H_2O, H_2, NH_3 and CH_4. Indeed, amino acids and other important components of living matter were produced experimentally in such a mixture of gases, under the action of electric discharges or by irradiation with suitable UV radiation.

5.5.4. METEORITIC COMPLEX

The meteoritic complex (Section 3.2(b) and 4.4.1) in the planetary system is also a secondary phenomenon. Its particles, ranging from submicron meteoroids to small asteroids cannot be considered to be original remnants of the parent protoplanetary disk. This is so because the particles of the complex have lifetimes considerably shorter than the age of the planetary system, e.g. $\sim 10^3$ yr for micron particles and

$\sim 10^5$ yr for cm particles. There are different destructive processes which limit their lifetime: the radiation pressure and Poynting-Robertson effect being best known ones.

The micrometeorites of submicron size are blown out from the planetary system by pressure of solar radiation. The radiation pressure on smallest members of the meteoritic complex namely exceeds the gravitational attraction. The larger particles reradiate the absorbed solar energy in all directions. This results in a progressive decrease of the angular momentum of the particles orbiting around the Sun. This effect (called Poynting-Robertson effect) causes a reduction in the eccentricity and major axis until the particle spirals near the Sun and sublimates. The sublimation of meteoroids (meteorites) becomes important at a distance of a few millions of kilometers. Closer to the Sun particles of the meteoritic complex are completely eliminated.

The collisions between meteoroids lead to their mutual erosion or fragmentation. Also the solar wind causes a sputtering erosion of the meteoroid surface, of about $\frac{1}{2} \mathring{A} \, yr^{-1}$. Capture and accretion of meteoroids by planets and satellites represent another loss mechanism.

All the mentioned processes (and some other less important ones) deplete the meteoritic complex, transferring its mass on the Sun, on planets and satellites or sweeping it out of the planetary system. The radioactive atoms in the meteorites fallen on the Earth indicate that the meteoritic material solidified more than 4×10^9 yr ago and that it broke up into small pieces at various times, less than 10^9 yr ago. From the crystal structure of the Ni–Fe meteorites the size of their parent body can be determined; apparently it was not more than 300 km in diameter.

5.5.5. ASTEROIDS, COMETS AND METEORITIC COMPLEX

The asteroids are probably made of the primeval matter of the protoplanetary cloud. They have been less affected by evolution than planets and Moon. It seems that the largest asteroids are original formations, while the small ones are collisional fragments deriving from a few primeval asteroids (about ten). Also a major part of the large meteoroids are probably collisional debris of the primeval asteroids (with high inclinations and large eccentricities). Comets, on the other side, are probably an important source of micrometeoroids and carbonaceous chondrites.

The meteoritic complex is a continuously renewed system. Losses are compensated for by pulverization and fragmentation of asteroids as well as by dust from nuclei of comets. A mass equilibrium is thus maintained.

There is an interrelation of comets with asteroids, in the sense that some Mars-crossing asteroids (such as Icarus) may be extinct nuclei of short-periodic comets. Their active period – when they emitted gases to form coma, tail and halo – depended on the rate of evaporation and the size of nucleus; it is estimated to last less than 10^4 yr. In their extinct period the cometary nuclei apparently stay much longer (10^7 to 10^8 yr). Duration of the latter period (called also the dynamical lifetime) is limited by collisions with or by close approach to the planets Earth, Venus, Mercury or Mars. As the extinct period is about 10^4 times longer than the active period, there should be about 10^4 extinct comets for each active short-periodic comet.

All processes in the whole solar system in general and in the meteoritic complex in particular prove that the evolution is far from being at its end. Matter, energy, momentum and angular momentum are being lost from, accumulated by or exchanged between members of the solar system. Analyzing the evolution of the solar system one realizes the fundamental role of the four interactions. But none of all the evolutionary processes can be considered to be specific for the solar system only. Any one could occur also in other, extrasolar planetary systems.

5.5.6. EXTRASOLAR PLANETARY SYSTEMS

According to a general opinion, many extrasolar planetary systems should exist in our Galaxy. No reason is known why a planetary system should have developed only for one of the 10^{11} stars of the Galaxy, many of which are indistinguishable from the Sun. The theory of the origin of our solar system indicates that formation of planets should be a common byproduct of formation of stars. The possible existence of planets orbiting other stars has deep implications for several disciplines (astronomy, exobiology, philosophy) and is a subject of general public interest.

There are indirect indications for existence of planetary systems around other stars. One is stellar rotation. The stars on the main sequence which are more massive than those of spectral type F5 rotate fast. On the contrary less massive stars rotate slowly. As the other properties (like color, temperature, spectrum, mass, chemical composition) change continuously, the conspicuous change of rotation in the F5 stars is explained by a phenomenon not directly observable: the stars cooler than F5 have planets with large angular momentum. If the angular momentum of the planets were concentrated into the central star, then its rotation would be much faster (thus the equatorial velocity of our Sun would be about $100 \mathrm{~km~s}^{-1}$ instead of the present $2 \mathrm{~km~s}^{-1}$; the angular momentum for the Sun is $1.6 \times 10^{48} \mathrm{~g~cm}^2 \mathrm{~s}^{-1}$ whereas the total momentum of the planetary system is $3.1 \times 10^{50} \mathrm{~g~cm}^2 \mathrm{~s}^{-1}$). If this argument is not wrong, the number of planetary systems in our Galaxy would be of the order of magnitude 10^{10}.

Astrometrists have tried to detect extrasolar planets during the last four decades. Their method uses the fact that an observed star and its unseen companion orbit around a common center of mass. Hence the position of the observed star should periodically deviate from a mean position. The amplitude of the oscillations is, however, very small (due to the small planetary mass) and very difficult to measure. The angular amplitude $<0.05''$ represents microns in focal plane of a large refractor. Possible planetary companions have been found by this very laborious method orbiting around some near stars (e.g. 61 Cygni – mass $0.008 ~M_{\odot}$; Proxima Centauri – $0.0018 ~M_{\odot}$; Barnard star – $0.001 ~M_{\odot}$). The results are rather uncertain and sometimes even confusing.

Other methods have been proposed for detection of extrasolar planets: spectroscopic observations of the periodic motion of the luminous primary star; large space telescopes; long-base radio interferometry, etc. Looking for intelligent electromagnetic signals from other galactic communities, for their intelligent machine probes or other non man-made artifacts is a quite different approach. If successful, it would prove existence of extrasolar planetary systems and of highly developed galactic communities as well.

5.6. Universe and Man

5.6.1. UNIVERSE AND LIFE

According to the present knowledge, the evolution of the universe culminated with the appearance of life – at least on the third planet in the solar system. The first evidence of life on the Earth is found in Pre-Cambrian rocks about 3×10^9 yr old. But more organized organisms (e.g., jelly-fish) appeared much later, about 650×10^6 yr ago.

(a) *Prebiotic Synthesis*

According to knowledge gained by astronomy, chemistry, biology and geology, life originated by a slow chemical evolution. The precursors of complex organic molecules have been probably synthetized already in the interstellar space. Molecules as complex as acetaldehyde CH_3CHO discovered by radioastronomers (Sections 2.2.4(f) and 4.3.4) exist in those places in our Galaxy where stars and eventually planetary systems are formed. The molecules help to cool the contracting nebula, accelerating thus its transition to the stage of protostar and planetary disk. Three of the molecules (i.e., formaldehyde H_2CO, hydrogen cyanide HCN and cyano-acetylene HC_3N) are important for the synthesis of amino-acids. The molecules of amino acids discovered in the Murchison meteorite are probably primordial molecules: they were formed in the parent interstellar cloud from which the solar system originated.

Complex organic molecules were apparently produced in the prebiotic phase of the Earth's evolution. Laboratory experiments imitating the conditions on the primitive Earth contributed substantially to the study of the chemical origin of life. Into a gas mixture corresponding to the primitive atmosphere energy was fed from different sources: UV radiation simulating solar radiation, prolonged 60 000 V discharges simulating lightnings, heat (simulating volcanoes), cosmic rays and radioactivity. All of them worked to some extent. The experiments led to the conclusion that all the organic molecules essential for life could have been produced in the primitive atmosphere of the Earth.

Concentration of organic molecules, formation of cells, evolution of multicellular organisms up to man have been explained (though with many large gaps) by biochemistry, biology, paleontology and anthropology. The life is thus conceived as a natural link in the evolution of the universe. If that is the case, life could be a frequent phenomenon in the universe.

(b) *Plurality of Inhabited Worlds*

Life in extrasolar planetary systems is possible and even probable. That is all that one can say at the present day, in the absence of a direct observational proof. Further considerations about forms of life elsewhere in the universe, the degree of its evolution, about intelligent galactic communities and communication with them through interstellar space or even their visits to the solar system, are at present no more than mere speculation.

(c) *Man in the Universe*

The human body is built up of 10^{28} to 10^{29} elementary particles interacting strongly in atomic nuclei and electromagnetically in atoms, molecules and tissues. Its mass (10^4 to 10^5 g) is 10^{28} to 10^{29} times smaller than the mass of a normal star like our Sun. As for mass, our body is quite negligible in comparison with the Sun, the nearest star and most important cosmic body besides our Earth. The medieval Italian saint Francesco d'Assisi called the Sun 'messer lo frate Sole'. Three quarters of a millenium later, we could scarcely find a better characterization of relation between man and the Sun.

5.6.2. SUN AND MAN

The Sun is man's brother in the sense that they are both descendents of the same primitive nebula. The atoms of the Sun and of our bodies were formed in ancient stars more than 5×10^9 yr ago, were ejected into the interstellar space, became part of the presolar primitive nebula; by condensation they went either into the Sun or into planets; by photosynthesis they entered the biosphere and by food into our organism.

As for energy, the Sun is our lord ('messer') and of the whole biosphere. The energy of solar radiation is stored in organic molecules (sugar, protein, fat) produced in green plants. The living organisms use the chemical energy to carry out the synthesis of more of themselves. By oxidation the energy is released and turned over into heat and work at constant temperature. Man is a constant temperature energy converter. An average man lying quietly in a warm room converts about 43 kcal h^{-1} (basic metabolism rate). It is the energy from the deep solar interior. Besides maintaining the energy flux through the biosphere, the solar radiation controls the environmental conditions on our planet.

Practically all the energy needs of modern society are covered from coal, oil or natural gas. These fossil fuels store the solar energy deposited in ancient times. Their amount is limited, while the energy consumption per person grows steadily. There are alternative energy sources, but the solar radiation is the only clean and inexhaustible energy source available for the future generations.

Besides the absolute dependence of all terrestrial life on the solar energy, there are secondary influences of the Sun on the Earth. The variations of the solar radiation (X-ray, UV, corpuscular, radio) accompanying solar activity, have repercussions in terrestrial magnetosphere, ionosphere, atmospheric circulation and biosphere. Among others, different functions of the human organism are influenced by solar phenomena, in particular by flares (helio-biology). This is a result of many statistical investigations. The mechanism of the influence is not yet quite explained, but the electromagnetic forces play an important role. By its activity the Sun influences our electromagnetic environment in the form of geomagnetic variations, increase of intensity of very long radio waves (atmospherics), electrotelluric currents in the soil, changes of vertical electrostatic potential and substantial increase of solar radio noise lasting from a few seconds to a few hours. On the other side, the processes in man's organism are mainly electromagnetic, for example: chemical reactions, electrostatic potential on cellular membrane, regulation of the heart rhythm, propagation of pulses along nerves, etc. The effects of the electromagnetic environmental changes may be beneficial, harmful or even lethal.

Besides the dependence of the biosphere in general and of man in particular upon the Sun and its activity, there is another important cosmic influence, viz. the cosmic radiation. Its flux into the atmosphere $(\sim 9 \times 10^4 \, erg \, cm^{-2} \, yr^{-1})$ represents in troposphere X-ray, γ-ray and corpuscular radiations equivalent of $0.03 \, R \, yr^{-1}$.

An organism becomes sensitive to the radiation effects after it has been exposed to more than 100 R during several days of irradiation. The dose of 500 R absorbed in a few days becomes lethal for 50% of the vertebrates; the other 50% become sterile. Such irradiation and even higher may have been produced by a supernova at a distance of 100 parsec or closer to the Sun. Statistics of supernovae indicate that during 10^8 yr the dose ≥ 200 R should occur about eight times, radiation exposure ≥ 500 R about twice, and irradiation ≥ 1000 R about once. The high irradiation could have caused the extinction of some species known to paleontology.

5.6.3. CONCLUSION

Astronomy with physics (high energy physics, nuclear physics, atomic physics, radiophysics) chemistry, biology, geophysics, paleontology,... present a coherent picture of our Universe in space and time. The hierarchy of structures and the evolutionary sequence culminate with living matter, in particular with the enormously complicated system of human organism.

Though very complicated, the human organism is part of the universe and does not violate its unity. With its 10^{28} to 10^{29} constituent elementary particles the man's organism remains 10^{28} to 10^{29} times smaller than a common star like our Sun. The difference in energy is still larger, about 33 orders of magnitude. The energy of the Sun is 3×10^{48} erg (kinetic energy of particles 2.7×10^{48} erg, radiant energy 2.8×10^{47} erg) and an average man has about 3×10^{15} erg of stored energy in his body.

Though quite negligible when compared with the mass and energy of cosmic bodies and with the dimensions of the universe, the man's mind is capable of understanding the universe, which is immense in space and time. Or expressed in the words of Blaise Pascal: 'L'homme est infiniment petit par son corps mais il est infiniment grand par son esprit'.

LIST OF SYMBOLS

Elementary Particles

γ	photon
e^-	electron
e^+	positron
ν	neutrino
μ	muon, μ-meson
π	pion, π-meson
K	kaon, K-meson
p	proton
n	neutron
$\Lambda, \Sigma, \Xi, \Omega$	hyperons
W	intermediate boson
N	baryon number
l	lepton number
s	spin
S	strangeness
$\sim$	(tilde) means antiparticle, e.g. $\tilde{p}$

Atomic Nucleus

A	mass number
Z	atomic number
$N(A, Z)$	nucleus
^{1}H	nucleus of hydrogen, proton
α	nucleus of helium, ^{4}He
q	electric charge ($+Ze$)

Mass and Density

m_e	rest mass of electron 9.109558×10^{-28} g
m_p	rest mass of proton 1.672661×10^{-24} g
g	gram: kg $= 10^3$ g, $t = 10^6$ g
M_E	Earth mass 5.976×10^{27} g
$M_\odot$	Sun mass $1.989(2) \times 10^{33}$ g
M_r	mass of stellar interior of radius r
ϱ	density
ϱ_m	density of matter
ϱ_r	density of radiation
n_e	electron density

Length and Dimensions

L^*	Planck length 1.6×10^{-33} cm
f	fermi; $1 f = 10^{-5}$ Å $= 10^{-13}$ cm $= 10^{-15}$ m
Å	Ångström; 1 Å $= 10^{-8}$ cm $= 10^{-10}$ m $= 10^5$ f
μ, μm	micron, micrometer 10^{-6} m $= 10^{-4}$ cm $= 10^4$ Å
r_g	radius of gyration, Larmor radius
R_g	Schwarzschild radius

$R_\odot$ Sun radius 6.9599×10^{10} cm
AU   astronomical unit (mean Sun – Earth distance) 1.495979×10^8 km = 1.495979×10^{13} cm
ly   light year 9.460530×10^{17} cm = 63240 AU
pc   parsec 3.0857×10^{18} cm = 3.2616 ly = 206265 AU
kpc   kiloparsec 10^3 pc
Mpc   megaparsec 10^6 pc
$R(t)$ expansion function
λ wavelength
λ_D radius of Debye sphere
ds distance of two neighbouring events; metrics

Time and Frequency

ms   millisecond 10^{-3} s
s   second
h   hour
d   day; 1 d = 86400 s
mo   month
yr   year; 1 yr = 365.24219 d = 31 556 926 s $\sim 3 \times 10^7$ s
H^{-1} Hubble time, expansion age of the universe
Hz   Hertz, cycle per second
ν frequency of electromagnetic radiation
ω_p plasma frequency

Energy, Power and Flux

eV   electron-volt 1.602×10^{-12} erg; temperature associated with 1 eV is 11 604.8 K.
$m_e c^2$ rest energy of electron 8.18727×10^{-7} erg = 0.511 MeV
$h\nu$ photon energy
erg 0.624145×10^{12} eV
J   Joule 10^7 erg
L_r luminosity of stellar interior within radius r
L   luminosity of a star or galaxy
L_x X-ray luminosity
$L_\odot$ Sun's luminosity 3.826×10^{33} erg s^{-1}
f.u.   flux unit 10^{-23} erg cm^{-2} s^{-1} Hz^{-1} = 10^{-26} W m^{-2} Hz^{-1}
Jy   Jansky; 1 Jy = 1 f.u.

Temperature

K   kelvin, the same as degree kelvin (°K)
T_b brightness temperature
T_e effective temperature
T_r temperature at distance r in stellar interior

Constants

c velocity of light in vacuum 2.997925×10^{10} cm s^{-1}
e elementary electric charge 4.803250×10^{-10} esu = 1.60219×10^{-19} coulomb
G gravitation constant 6.67×10^{-8} dyn cm^2 g^{-2}
h Planck constant 6.6262×10^{-27} erg s
$\hbar$ rationalized Planck constant 1.05459×10^{-27} erg s
H Hubble constant (value not well known)
k Bolzmann constant 1.38062×10^{-16} erg K^{-1}
R gas constant 8.31434×10^7 erg K^{-1} mole^{-1}

Mathematical signs

∇	operator nabla, del
$\propto$	proportional to
$<$	smaller than
$\ll$	much smaller than
$\sim$	roughly equal
$\approx$	approximately equal
$\Rightarrow$	sign of implication
Δ	sign for increment or small difference (e.g. Δr, Δm)

Various

G	gauss; $1\,G = 10^5\,\gamma$
P	pressure
sr	steradian; $1\,sr = 3282.8\,deg^2$
ε	energy production per gram
κ	opacity of stellar material

INDEX OF SUBJECTS

GEOPHYSICS AND ASTROPHYSICS MONOGRAPHS

AN INTERNATIONAL SERIES OF FUNDAMENTAL TEXTBOOKS